Land Restoration and Reclamation

Land Restoration and Reclamation: Principles and Practice

James A. Harris, Paul Birch and
John P. Palmer

Longman

Addison Wesley Longman
Addison Wesley Longman Ltd,
Longman House, Burnt Mill, Harlow,
Essex CM20 2JE, England
and associated companies throughout the world

First published 1996

British Library Cataloguing in Publication Data
A catalogue entry for this title is available from the British Library

ISBN 0-582-243130

Library of Congress Cataloging-in-Publication Data
A catalog entry for this title is available from the Library of Congress

Typeset by 16 in Times 10/12pt
Produced by Longman Singapore Publishers (Pte) Ltd.
Printed in Singapore

Contents

Foreword

Land reclamation has always been an unwanted stepchild of ecologists. The thralldom of the Galapagos, a pristine cloud forest or a coral reef have traditionally drawn the attention of biologists and the hearts and wallets of environmental preservationists. Only a few die-hards of the ecological world have dared to investigate damaged landscapes – such as Tony Bradshaw in the United Kingdom and John Cairns in the United States. Until recently, only a handful of serious scientists dedicated their careers towards restoring ecosystems, such as the late Ted Sperry who returned cropland to North American prairie.

But our entire planet consists of damaged landscapes, with only here and there a parcel of Eden remaining. As our global population expands, we pay an increasingly dear price for the free environmental services we once took for granted, such as potable water, fertile soil, floodwater detention, expansive recreational lands and abundant timber. As the gloom of this reality settles upon us, we contemplate how important the reclamation of damaged landscapes is to our economy. The likes of Bradshaw, Cairns and Sperry suddenly draw our attention.

Previously, 'reclamation' had merely meant draining wetlands or recouping mined lands for economic purposes. By the later 1970s, experimentation was well under way to restore natural ecosystems on reclaimed land, spurred in part by legal requirements for compensatory mitigation, but fostered more by a new yearning in our culture. It was as if our collective unconscious told us we should restore Nature. The restoration movement was under way – an idea whose time had come. Everyone seems to be participating – public resource agencies, non-governmental organizations, industrial interests, environmental consultants, ecotourists and weekend volunteers.

Everyone, that is, except the academic community. University ecologists continue to traipse off to the Galapagos and lament the demise of Nature. There may be several reasons for their resistance to restoration ecology. Perhaps the multivariate conditions at restoration sites seem forbidding to researchers desiring neat experimental settings. Or possibly the division of university budgets into narrowly conceived academic departments has deterred needed interdisciplinary collaboration. Then again, restoration may demand longer-term research than professors can afford, if they are ever to become tenured. Whatever the reason, curricula in restoration have largely been relegated to *ad hoc* seminar courses, and textbooks have remained unwritten.

That is until now. The authors of this book have produced a text to serve a new curriculum in restoration ecology being introduced at the University of East London. This text marks the entrance of the academic sector into its rightful place in the restoration movement. For that reason, this book assumes historic importance, as well as contemporary relevance.

Andre F. Clewell, Quincy, Florida USA
President, Society for Ecological Restoration

Preface

The topic of land reclamation and restoration is becoming an increasingly important focus of activity for scientists, professionals and practitioners in a range of disciplines, from engineering through archaeology and biology to the social sciences. There are many reasons for this: the subject lends itself inevitably to interdisciplinary approaches, indeed without this restorations and reclamations will fail; the pressure on the use of land has never been greater; the realisation that the conservation of existing ecosystems is simply not enough to ensure our own species' future; and the task of land restoration is the ultimate test of our understanding of the long-term way in which ecosystems work.

The aim of this book is to provide a text which takes the reader, be they undergraduate or postgraduate, practitioner, professional or community member, through the processes which are required to ensure the long-term sustainable reuse of land. We believe this to be a unique text both in approach and content, and estimate that it will provide an aid for those teaching the related courses appearing in our universities and colleges.

It is in the nature of our industrial legacy that we have to start at a less than optimum place in the land use change cycle, dealing with damage, rather than protecting the land from it. The mechanisms underlying the way in which ecosystems work and how the factors of stress and disturbance impact upon them are outlined. These conditions are important in setting the theoretical framework of understanding to enable a systematic and quantifiable approach to producing sustainable outcomes for the future. A description of the actual processes of degradation, with examples from a variety of degradative land uses follows. The importance of adequate assessment, and a consideration of the treatment of green field sites is included. The text then enters the more familiar ground often considered by texts on the topic of land restoration and reclamation; the treatment of degraded systems. Finally, we consider how failures occur, and outline issues which must be addressed in the future to prevent such failures and improve practice.

It is intended that this text should act as a framework for discussion of the major processes, issues and concepts involved in the process of land restoration, and point to further sources for consultation for wider and deeper exploration.

J.A.H., P.B, J.P.P.

Acknowledgements

J.A.H. would like to thank E.C. Harris for help in the preparation of the index.

We are grateful to the following for permission to reproduce copyright material:

Fig. 1.1 reproduced from DeAngelis, D.L. (1992) by permission of D.L. DeAngelis. Fig. 2.2 reproduced from Davidson, D.A. (1992) and table 3.3 reproduced from O'Riordan, T. (1995) by permission of Addison Wesley Longman Limited. Fig. 5.3 reproduced from Bradshaw, A.D. and Chadwick, M.J. (1980), fig. 5.5 reproduced from Chadwick, M. J. and Goodman, G.T. (1975), tables 1.2 and 1.3 reproduced from Bradshaw, A.D. and Chadwick, M.J. (1980), table 3.2 reproduced from Ramsay, W.J.H. (1986) and table 6.2 reproduced from Begon, M., Harper, J.L. and Townsend, C.R. (1990) by permission of Blackwell Science Ltd. Fig. 6.3 reproduced from Hall, I.G. (1957), table 6.2 reproduced from Bradshaw, A.D. (1994), tables 6.3 and 6.4 reproduced from Mars, R.H. and Bradshaw, A.D. (1993) by permission of the British Ecological Society. Table 2.5 reproduced from Spellerberg, I.F. (1991) by permission of Cambridge University Press. Fig. 3.7 reproduced from Howsam, P. (1990), tables 5.1, 6.10 and 6.11 reproduced from Gilbert, O.L. (1991) and table 6.12 reproduced from Usher, M.B. (1986) by permission of Chapman & Hall. Box 5.15, figs 5.15, 5.16 and 6.1, and table 5.10 reproduced from Coppin, N.J. and Richards, E.G. (1990) by permission of the Construction Industry Research and Information Association, London. Figs 4.1, 4.4, 4.5, 4.6, 4.8, 4.9, 4.10, 4.11 and 5.4 reproduced from Richards, I.G., Palmer, J.P. and Barratt, P.A. (1993) and fig. 5.19 reproduced from Edgerton, D., Harris, J.A., Birch, P. and Bullock, P. (1995) by permission of Elsevier Science Ltd. Fig. 1.5 reproduced from Department of the Environment (1991) by permssion of the Department of the Environment. Fig. 3.1 reproduced from Moffat, A. and McNeil, J. (1994) by permission of the Forestry Commission. Crown Copyright. Crown copyright is reproduced with the permission of the Controller of HMSO. Top back cover picture reproduced by permission of Jeanetta Ho. ©1990 Jeanetta Ho. Courtesy of Cleveland Metroparks, Brecksville, Ohio, USA. Table 6.1 reproduced from Huby, M. (1981) by permission of Dr Meg Huby. Table 5.6 reproduced from Coppin, N.J. and Bradshaw, A.D. (1982) by permission of Mining Journal Books Ltd,

London. Fig. 6.5 reproduced from Petit, D. (1982) by permission of Prof. D. Petit. Figs 6.11 and 6.12 and table 6.14 reproduced from National Rivers Authority (1994) by permission of the Royal Society for the Protection of Birds. Fig. 5.13 reproduced from Shildrick, J. (1984) by permission of the Sports Turf Research Institute, Bingley, UK. Table 6.16 reproduced from Scullion, J. (1994) by permission of Dr J. Scullion. Figs 4.2, 4.3, 5.7, 6.14 and 6.15 and tables 4.1, 4.4, 5.8 and 6.18 reproduced from Welsh Development Agency (1993) by permission of the Welsh Office and the Controller, HMSO.

Whilst every effort has been made to trace the owners of copyright material, in a few cases this has proved impossible, and we take this opportunity to offer our apologies to any copyright holders whose rights we may have unwittingly infringed.

Causes of degradation and aims of restoration

Chapter 1

Land as a resource

1.1 Introduction

Land is one of our most precious resources. We need land for constructing our homes, industry and infrastructure, to provide a soil for our crops to grow in, to provide open space and wild places for our recreation, and to allow natural terrestrial ecosystems to thrive and secure genetic resource for the future. Water, minerals and energy for our very existence are obtained from land. We use land for storage of water and disposal of waste materials. Its vegetation allows us to harness the productivity of our domesticated and wild animals, and our crops and forests; whilst providing a green lung to clean our air. Humankind's utilization of land has led to all areas of the Earth being modified. It is difficult to identify significant areas of vegetation in the world which have not been so altered. Europe has seen some of the most intensive use of land globally for agriculture and industry, and contains some of the most densely populated countries of the world; of which the United Kingdom is one. Even in the United Kingdom, however, population density and extent and type of land utilization varies considerably between regions (Table 1.1).

Table 1.1 Population density and land utilization in the United Kingdom

Indicator	UK range
Population density	< 50 inhabitants/km^2 (North-east Scotland) to > 1000 inhabitants/km^2 (London, North-west England and West Midlands)[a]
Employment in agriculture	1.2% of total employment (South-east England) to 4.6% of total employment (South-west England)[a]
Employment in industry	25.4% of total employment (South-east England) to 39.1% (West Midlands)[a]
Employment in services	57.7% of total employment (West Midlands) to 72.5% of total employment (South-east England)[a]

[a] Source: European Commission (1994).

Increasingly sophisticated methods of land exploitation are being used as resources become depleted, and ever higher human demands inevitably lead to the degradation of land as a resource. Realization that such degradation may become damaging, not only to people using or living near to such land but also to future generations, has led governments to adopt policies that favour activities resulting in more sustainable land-uses. The making safe and productive reuse of degraded land is central to achieving sustainable land-use, and is the principal theme of this book. We will be considering the effects of anthropogenic activities on natural and semi-natural systems, and the need to restore their self-sustainability. We will also consider the need to alter land required for other end-uses, such as building, where the site is not in a state where such activities could be carried out with long-term safety, in the context of sustainability of use of all the land.

Central to our considerations, therefore, is a definition of what is meant by 'sustainability'. There are many definitions of this term, and many more interpretations. One strict dictionary definition is 'to maintain or prolong', another 'to provide for or give support to, especially by supplying necessities' (*Collins English Dictionary*, 1979). Therefore the concept has evolved that the sustainable use of the land is one which can be carried out in perpetuity. This is achieved by keeping the land in good order by supplying the physical, chemical and biological inputs required for its long-term maintenance, which may be in the form of (for example) water resources, mineral nutrients, or specific cultivations such as mowing. It does not exclude development of human activities within a landscape; indeed, it is impossible to consider any system, dependent on scale, which does not include humankind's activities, as humanity is not divisible from nature. The definitions of sustainability have been further extended recently to include the involvement of specific cultural attributes such as learning and networking in developing an accurate assessment of the true 'worth' or 'value' of land (Meadows *et al.*, 1992). A sub-set of this, and a term which we shall also use, is 'self-sustainability', essentially a system which requires no input from humankind to be maintained in the long term, such as you might expect of an undisturbed ecosystem. Whether any such ecosystem currently exists is a matter for discussion elsewhere.

It follows, therefore, that in achieving sustainable land-use we must consider the question of scale – in both size and time. Many of the adverse changes brought about on a particular area of land can be reversed if there is compensation elsewhere in the system. This compensation may take the form of restoring a wildlife system whilst allowing an economic development to proceed (if it is of minimum environmental impact) which finances the protection of the wildlife area. Compensation for loss of hydrological flexibility in one place may be achieved by protecting floodplain function further up or downstream. We must also bear in mind the *precautionary principle*, that is because of the shortage of data, deficiencies in our models, and the possibility that some things are beyond knowing. O'Riordan (1995) has outlined the four meanings of the precautionary principle:

1. Thoughtful action in advance of scientific proof; in other words better to act now rather than when it is too late.
2. Leaving ecological space as room for ignorance; not extracting resources to their limits of exhaustion, when we are unable to predict the consequence of this action.
3. Care in management; participation of society at a number of levels in the trial and error involved in development and change of land-use.
4. Shifting the burden of proof from the victim to the developer; this is controversial, but how much risk is acceptable in an unknown form by a development process must be taken at the societal level, without prejudice to sustainable growth.

The reclamation and restoration of land has a major part to play in satisfying the first two meanings of the precautionary principle and will be impacted upon by the last two.

The first matter we will consider is how terrestrial ecosystems function and how they have developed. These factors are of central importance, as without an understanding of how things have to be the way they are and how they function, there is little prospect for achieving restoration that works.

1.2   Function of terrestrial ecosystems, nutrient cycling, soil structure, biological assemblages and interactions

1.2.1   Ecosystem function

Ecosystems maintain their stability as a result of a complex interaction of production, consumption and cycling, of gases, solutes and liquids (Fig. 1.1). The complex interactions of biological material and mineral material are termed *biogeochemistry*. This discipline has gained much attention in recent years in Western science, although it has a history of a century of study in Central and Eastern Europe.

The driving force of ecosystem function is the fixing of solar energy into biological molecules by *photoautotrophs* for later use, or other organisms which may consume the photoautotroph. This process leads to a rise of *free energy* in the system, allowing work to be done. Photoautotrophs are principally green plants in terrestrial systems, and algae in the oceans.

1.2.2   Functional groups

For an ecosystem to be fully functional (and therefore self-sustaining) there must be representatives from the three metabolic groups: *primary producers, consumers* and *decomposers* (Stolz *et al.*, 1989).

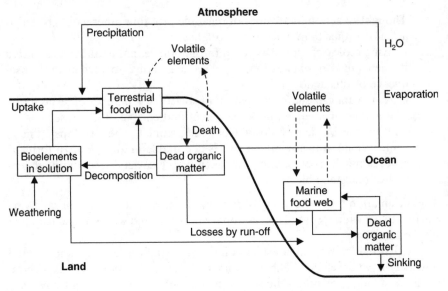

Fig. 1.1   Generalized biogeochemical cycle (from DeAngelis, 1992).

Primary producers   These are the organisms that absorb solar radiation and fix it in organic molecules by the process of *photosynthesis*. Net primary production is the total energy fixed less that used by the primary producer in respiration. For the majority of terrestrial systems the primary producers are green plants. These plants produce carbohydrates, in the first instance, which are then used to produce proteins, enzymes, fats and vitamins; required in varying degrees by other functional groups within the ecosystem. The construction of these complex molecules requires elements other than carbon and oxygen, supplied by the soil solution.

Consumers   These organisms which feed on the primary producers are usually animals eating plant parts, above and below ground. These consumers have an important function in dispersal of plant propagules and organic matter, and directly return carbon to the atmosphere in the form of carbon dioxide. Another form of consumer is the carnivore, which is a secondary consumer feeding on the herbivores.

Decomposers   These organisms break down the organic compounds in dead primary producers and consumers, returning elements to their mineral forms for recycling and reuse. Without their activities there would very quickly be an accumulation of organic matter on a global scale, exhausting the atmospheric pool. Decomposers also have subsidiary functions such as the development and maintenance of soil structural stability, by binding soil particles together as a result of producing gums and mucilages. They consist primarily of the bacteria, fungi and protozoa.

Functional systems may be very small, such as microbial mats just a few centimetres across, to very large such as tundra hundreds of kilometres wide. What is clear in all of them is that all functional groups need to be present in order for the system to be self-sustaining.

1.2.3 Carrying capacity, biodiversity, order and chaos

There are many characteristics of natural systems which may be affected by the activities of man. Principal amongst these are carrying capacity (K) and biodiversity. The *carrying capacity* of a system is the theoretical maximum density that it can sustain (Odum, 1993). Density can be applied with respect to individual species, community assemblages and total biomass (usually indicated by the amount of carbon associated with viable biomass) and represents the upper limit of the system. A combination of factors will control the size of the carrying capacity, namely solar radiation, climate, altitude, environmental stress, and frequency and amplitude of disturbance. The nature of these characteristics will depend upon soil type, aspect, topography and geographical location.

1.2.4 The importance of biodiversity

The majority of ecosystems on the planet Earth are open ones in thermodynamic terms, i.e. they exchange energy and materials with each other. Exchange occurs, principally, by interchange of gases and liquids between areas, but there can also be exchange of *biological potential*. The biological potential of a system may be expressed as the amount of genetic diversity and biomass able to do work and reproduce in a self-sustaining way. The biological potential is of critical importance in the recovery of systems after natural disasters, exposure of new substrates (such as the result of the retreat of glaciers or rises in temperature) and, for our purposes, the re-establishment of self-sustaining, functioning ecosystems in restorations and productive systems on reclamations.

This biological potential is usually referred to as the *biodiversity* of a system, and is reflected in the number and abundances of species of all types within a particular functional unit; such as a catchment or island (Wilson, 1992). The amount of genetic information in the system allows it to maintain its structure and function. Such maintenance is achieved (in the large majority of ecosystems) by fixing sunlight to produce free energy allowing the production of biomass, enabling work to be done, and updates and maintains stores of information, i.e. genetic material. If the system becomes disordered as a result of a perturbation, be it man-made or natural, then its *entropy* increases. Increase in entropy means there is more disorder in the system. Such disorder can take the form of an increased abundance of small molecules, such as nitrate ions, that were formerly bound in functioning biomass. Such changes in the ratio of large to small molecules occurs in systems which are stable over long timescales, but there is usually sufficient information retained in the system to allow it to recover when the perturbation subsides. If the pressure of perturbation is maintained, however, there comes a point where there is insufficient information to compensate for loss of order, and a *catastrophic collapse* occurs. Such a collapse may be directly represented as a loss of a biodiversity from a system and occurs when a vital function can no longer be fulfilled. Since the full potential of individual species to respond in an entropy-minimizing manner to environmental perturbation is unknown, it is essential, applying the precautionary principle, that individual species should be preserved (Lockwood and Pimm, 1994).

1.2.5 Soil and ecosystem development

The majority of terrestrial ecosystems are based on soil substrates. Soil is composed of mineral and organic matter, air, water and living organisms. The soil is a dynamic living system which is capable of supplying plants with all their requirements for growth, apart from solar radiation and carbon dioxide. In addition, the soil provides a means of physical support and anchoring for plants.

Soils are formed as a result of the interaction of mineral substrates, climate, organisms, topography and time. Most soil systems are derived initially from exposed bedrock, but peat soils are composed almost entirely from organic matter. In the case of peatlands, there may have been mineral soils which were replaced by the formation of peat. Weathering results in the mobilization of mineral materials from the rock, and at this point several processes can occur. Loose material may be washed or blown away, depending upon the topography, exposure and climatology, in which case the loose materials will eventually settle somewhere to produce *loess* soils, i.e. sandy materials blown in, or may settle out downstream in river systems, so-called *alluvial* soils. Glaciers can grind bedrock to a fine till, which upon exposure makes an ideal material for the establishment of plant and algal pioneers. Finally, in many cases pioneer organisms such as lichens may take hold. These are symbiogenetic organisms which consist of an interaction between a fungus, which provides anchorage and protection, and an algae (true algae or blue-green) which can fix carbon by photosynthesis. As a result of their activity the lichen produces organic acids, accelerating the weathering process.

The weathering process results in the production of a material which has an organic content and a developing microbial community, allowing the establishment of more complex plants requiring nutrients to be released from the mineral particles or fixed from the atmosphere. Nitrogen fixation into the soil system is of central importance in the development of the biological community, as it is not found in a plant-available form in bedrock or mineral materials. The usual route of succession is as shown in Fig. 1.2. There are some cases where, for climatological or topographical reasons, climax forest or peat bog is not established, such as sub-arctic tundras or desert valleys.

1.2.6 Nutrient cycling

Nutrient cycling is of central importance in any ecosystem, since without it there would be accumulation of nutrients in a form unavailable for use by living organisms.

Carbon cycle The rate of fixation by green plants is such that without recycling atmospheric carbon would be exhausted within 20 years. Consumers and decomposers become important here – they break down the carbohydrates and other compounds synthesized by plants and incorporate them into their own biomass, whilst returning carbon to the atmosphere in the form of carbon dioxide (Fig. 1.3). There are some instances where carbon compounds can form an obstacle to reclamation or restoration, i.e. when they are present in toxic forms, usually as a direct result or as by-products of industrial processes. In these cases they need to be returned to harmless or useful forms, or removed.

Nitrogen cycle As mentioned previously, nitrogen is almost entirely absent from all but carbonaceous bedrock and must be fixed from the atmosphere.

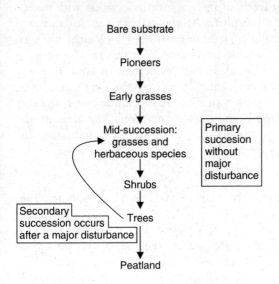

Fig. 1.2 Primary and secondary succession.

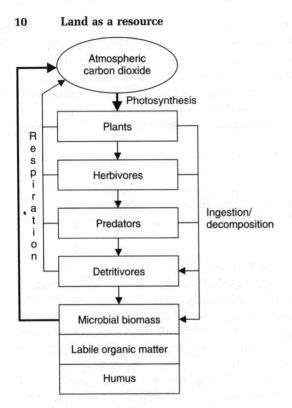

Fig. 1.3 The carbon cycle.

Certain micro-organisms, living freely or in symbiotic association with plants, are able to fix nitrogen from the atmosphere. Although there is some fixation of nitrogen into oxides by the action of lightning, without biological nitrogen fixation these systems would be unable to develop. For this reason nitrogen plays an important role in the reclamation and restoration of land. In many substrates nitrogen needs to be added, initially or over the long term, or its fixation encouraged in order to maintain an effective vegetation cover. In other situations nitrogen is in over-supply, and must be removed. The amount and form of nitrogen in the system will play a large part in determining the composition of the vegetational assemblages of terrestrial ecosystems.

Other mineral cycles   Other minerals, such as potassium, phosphate, iron, calcium and copper, are also important in controlling the size and composition of ecosystem communities, and have their own peculiarities of cycling. Problems arise when such minerals are absent from the system, and sometimes when they are in over-abundance. Over-abundance of the heavy metals such as lead and zinc can be a significant difficulty, as will be demonstrated in subsequent chapters.

1.2.7 Soil composition

The principal features of soils which govern their *inherent* characteristics are their *particle size distribution* and *organic matter content*. Soil particles have a large range of sizes as a result of weathering, which has a direct consequence for their physical and chemical characteristics (Table 1.2). The three classes of material, sand, silt and clay, are used to delimit the *textural triangle* (Fig. 1.4). Particles greater in size than 2 mm have little impact on soil chemistry. The classes that result from the distribution of these particles are a useful tool in predicting the general physical and chemical characteristics of the soil. Clayey soils will have high nutrient-retaining capacities, owing to their *cation exchange capacity* (CEC), i.e. their ability to retain cations, but will be very poorly draining, sticky, and tend to droughtiness and waterlogging. Sandy soils, although free-draining, have poor nutrient retention characteristics. Silty soils have moderate nutrient-retention and drainage characteristics, but the ideal balance is found in the loam soils. In addition to this basic set of characteristics are imposed the effects of organisms and organic matter. Organic material can greatly improve the nutrient-retaining characteristics of sandy soils, and the drainage characteristics of clay soils. Peat soils are inherently waterlogged, as otherwise they would not form, but when removed from their location, or if the water table is lowered, may become freely draining, and capable of retaining high amounts of nutrients.

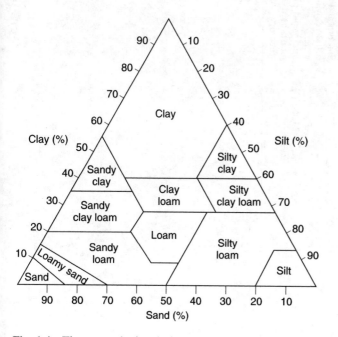

Fig. 1.4 The textural triangle for soil.

Table 1.2 Properties of soil particles dependent upon size (Bradshaw and Chadwick, 1980)

Type	Size (mm)	Number per gram	Surface area (cm²) per gram	Physical character	Mineralogy	Chemical characteristics
Coarse sand	0.2–2.0	5.4×10^2	21	Loose, neither sticky nor plastic, low water retention	Mainly quartz with some fragments of rock	Little or no ability to retain ions
Fine sand	0.02–0.2	5.4×10^5	210	As above	Quartz and feldspar, with some ferromagnesians	As above
Silt	0.002–0.02	5.4×10^8	2 100	Smooth and floury, slightly cohesive, will retain some water	As above with some mica and clay minerals	Some ability to retain cations
Clay	<0.002	5.4×10^{11}	23 000	Sticky and plastic when moist; hard and cohesive when dry	Clay minerals, such as bentonite	High ability to retain cations, and to adsorb onto organic complexes

It is not possible, however, simply to construct a soil from the ingredients of sand, silt, clay and organic material, and expect it to perform in the same way as a soil with the same characteristics in, say, a grassland that has developed over a long period of time. Such a grassland will have many interactions and interconnections between different species of organisms from different functional groups, an established bulk density, pH and stone content.

These characteristics will be found in the soil even if it is moved, although it is likely that the organic matter would change with time. The *in situ* characteristics are a feature of the interaction of the inherent characteristics with features of the site such as topography, climate, sources of biological potential and *land management* or *anthropogenic impacts*. When not impacted by humankind, soils will eventually reach an equilibrium with their biological communities and in the fullness of time be recycled yet again through the Earth's crust.

1.2.8   Plant growth

Plants require the following elements to grow successfully:

1. Macronutrients – carbon, oxygen, hydrogen, nitrogen, potassium and phosphorus.
2. Micronutrients – calcium, magnesium, sulphur, chlorine, iron, manganese, boron, zinc, copper and molybdenum.

Without these elements, deficiencies will occur. Different species and subspecies may have very different requirements, even in the same community. Nitrogen is the common limiting factor in most ecosystems, and this is usually so in land reclamations and restorations.

1.2.9   Nutrient status

Physicochemical characteristics vary from one soil type to another, but there is a range within which we might expect normal plant growth (Table 1.3). A task of any restoration programme will be to alter substrate conditions to fall broadly within these limits. The prescription in Table 1.3 is not, however, capable of being applied universally as some communities that we wish to encourage will require some quite specific conditions out of this range.

1.2.10   Systems in flux

When any of the characteristics outlined above are perturbed, the stability of the system will be affected and will not return until another equilibrium point is reached. Achieving equilibrium may occur quite naturally during the course of succession or may result from direct intervention by humans, who may have

Table 1.3 Normal range of soil characteristics in available form for plant growth (from Bradshaw and Chadwick, 1980)

Parameter	Range (low–high)
pH	5–7.5
CEC (meq/100 g)	10–30
Nutrients (p.p.m.)	
K	100–300
Ca	500–2000
Mg	50–300
Fe	5–200
P	5–20
NH_4–N	2–20
NO_3–N	2–20
Mineralizable N	50–200
Total N (%)	0.1–1

created the original perturbation. The terms which we use to define the processes leading to the degradation of systems and the means of overcoming such degradation also need definition.

When land is being reclaimed for purposes such as construction, then other factors, such as geotechnical stability and hydrology, need to be taken into consideration.

1.3 Definitions

The term 'degraded land' has already been used to denote land damaged in some way by human activities. Other terms such as despoiled, disturbed, devastated, derelict, disused, damaged, contaminated and polluted are also used and interchanged often without precise definitions being made. Some countries of Europe also do not recognize all of these categories and emphasize one more than another. So, for example, in France and Spain the terms '*friche industrielle*' and '*ruinas industriales*' are used for derelict land, but are not widely understood outside of the specialist groups dealing with these issues. On the other hand, the terms '*terrain contaminée*' and '*suelos contaminados*' for contaminated soils are much more widely understood. Similarly, terms used to describe means of dealing with degradation are often confused; words such as restoration, reclamation, rehabilitation, remediation and amelioration. The definitions used in this book may be considered under two categories, those used to define the *problem* and those used to define their *solution*. In this book the terms will be used often and with distinct meanings. There are many processes by which land can become degraded and there are other mechanisms by which we can return sites to self-sustaining or other uses. The definitions are as follows.

1.3.1 The problem

Disturbance This term may be used to describe a wide range of phenomena pertinent to the field of land restoration. In its strict ecological sense disturbance is any event which causes a sudden change in the nutrient status of the system (Grime, 1979). This may be a *destructive disturbance* in which part of the existing biomass is killed, thus providing nutrients for the surviving biomass; or an *enrichment disturbance*, in which the carrying capacity is increased by the addition of nutrients from outside of the system or by changing the physical environment (Harris and Birch, 1992). Nutrient addition could be as a result of a fertilization event, or some natural phenomenon such as seasonal flooding found on flushing meadows. In both, changes in the size of the standing crop and its species composition will result.

Disturbance has also been used in the more traditional sense: to disarrange or muddle, which is broad enough to encompass almost all of humankind's activities, and is therefore of limited utility. A better term for an event or sequence of events which does not result in the gross impacts defined above may be *perturbation*.

Stress This word may be used to describe environmental factors which prevent the accumulation of additional biomass, i.e. this is the limiting factor of a system, when in a state of equilibrium, but finds special use here when intending to describe a downwards pressure on the carrying capacity of a system. A salt marsh may receive a similar amount of solar radiation as a nearby grassland, but is constrained in the biomass of its standing crop because the primary producers of the system have to divert part of their metabolic energy to maintain osmotic conditions required for exchange of solutes and other membrane gradients. In systems subject to increasing deposition of lead, for example, there will be a reduction in the carrying capacity of the system.

The ways in which carrying capacity may be affected by stress and disturbance are shown in Fig. 1.5.

Degradation This is the combination of processes which leads to the land under consideration being no longer fit for a wide range of uses from natural systems to building sites. This may take the form of geotechnical instability resulting in subsidence or landslips, or as erosion by wind or rain. There may be uses to which such land may be put, but whenever a self-sustaining system is the end-point, then remedial action will be required.

Derelict land This is land so damaged by industrial or other development that it is incapable of beneficial use without treatment. This definition for derelict land is the one used by the United Kingdom Department of the Environment in its surveys of derelict land which are carried out periodically.

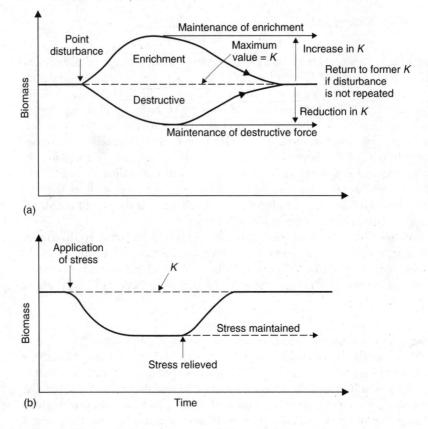

Fig. 1.5   Effects of (a) disturbance and (b) stress on biomass.

Contaminated land   This is land which contains substances that, when present in sufficient quantity or concentrations, are likely to cause harm directly or indirectly to humans, the environment or other targets such as construction materials. This definition of contaminated land is modified from one first adopted by the NATO Committee for Challenges to Modern Society and is useful because it clearly states that the concept of contamination is based not only on the presence of substances in the ground but also on their likelihood of causing harm to a particular target. Not all derelict land is contaminated and not all contaminated land is derelict. Both derelict and contaminated land may be disused and degraded.

1.3.2   The solution

Reclamation   This is a process by which previously unusable land is returned to a state whereby some use may be made of it. It usually refers to areas which were heavily contaminated or were geologically unstable, and may now be used for civil engineering, construction or limited growth such as sports fields which require high inputs to be maintained. It used to be the case in the United Kingdom about 20 years or more ago, that once the issues of site safety, provision of flat land for industry, and visual impact amelioration were dealt with, the process was considered finished. The use of the term reclamation nowadays is quite different and may incorporate conservation of features of wildlife, archaeological and mineralogical value. This is due to a change in perception about what is important. Reclamation may also be considered as the first stage of restoration, the process by which degraded soils are returned to a state where they may be prepared for growth, characterized by a lack of structure and nutrients, but also without any chemical impediment to growth such as the presence of inorganic contaminants. In most cases reclamation schemes will include an element of the area which is restored, and restoration principles will be incorporated, in the sense described later in this chapter, but cannot be defined as 'restored' as this cannot be applied to the whole area.

The term reclamation is also used for the process by which land is gained from submerged coastal lands, finding special application in the polder system of the Netherlands, and from desert and wetland areas. Such reclamation is in conflict with the concept of 'ecosystem restoration' which aims to conserve such systems as functioning units.

Rehabilitation   This term is often applied to areas which formerly had no growth at all, but with careful fertilization and landscaping works may be used to grow a limited number of plant species. Examples may be found again in sports fields, amenity grasslands and agro-forestry, but these systems are not self-sustaining, and the term is rarely used in the United Kingdom, and has no statutory basis.

Restoration   The definition of this term has given rise to much debate. Central to this debate has been the use of the term 'indigenous'. The 1994 definition adopted by the Society for Ecological Restoration is 'Ecological restoration is the process of repairing damage caused by humans to the diversity and dynamics of indigenous ecosystems'. This may be well characterized in North America, but there are few, if any, 'indigenous' systems left in Europe upon which to base our targets or end-points.

Another definition which is stricter in the ecological sense is 'Land restoration is the process by which an area is returned to its original state prior to degradation of any sort, i.e. back to a fully-functioning self-sustaining ecosystem', and refers to a specific area of land under consideration. This does not have to be a highly productive system; it could be, for example, the re-establishment of a salt marsh system on an estuary site or scrub land in arid desert. Whether or not true restoration can ever be fully achieved is a vexed question, because natural systems may change with time, depending upon which level of detail and at what scale the system is examined. The definition does not exclude the role of humankind in restoration as it must be recognized that humans form part of almost all self-sustaining ecosystems, including the indigenous ones. In many cases without constant, judicial, human intervention, 'restored' systems will fail. Without the stewardship of the native peoples of Northern America the European settlers would not have discovered a highly productive landscape, rich in game and vegetation, a situation which is repeated globally (McNeely, 1994). It does tend to tie us to a narrow end-target, i.e. reproducing the system present prior to disturbance. This may not be a bad thing, but it must be recognized that on the wider scale it may be of positive benefit to create a new ecosystem, different but more 'valuable' than the one present prior to the perturbation, which is a worthy reclamation end-goal.

Some have argued that it is impossible to restore degraded natural habitats. Gunn (1991), however, has clearly argued that provided that species have not been made extinct as a result of the degradation, then restoration is possible. If restoration is used in an absolute sense, as being *identical* to what has gone before, then the word has no practical utility.

Finally the use of exotic (non-indigenous) species in some circumstances must not be ruled out. There may be cases where such species may fulfil a role formerly carried out by a species which has become locally or globally extinct.

End-uses   There are two terms which are commonly used for describing the outcome of reclamation or restoration programmes: 'soft' and 'hard' end-uses.

Soft end-use applications require varying degrees of bioengineering and ecosystem re-establishment, in that plants form a major component of the final use. These may be further subdivided into productive and amenity end-uses:

1. Productive – arable agriculture, grassland agriculture, forestry, energy plantations, and glasshouse crops and horticulture.
2. Amenity – country parks, nature reserves, educational areas, campsites, golf courses, urban parks and 'landscaped' areas forming part of hard end-use developments.

These may have varying degrees of conservation value.

Hard end-uses have a high degree of engineered content and may contain no living component, apart from humans and their commensals and parasites: industry, reservoirs, housing, playgrounds, roads and car-parks, commercial development and public sector building.

As to the balance between the two, this is a societal decision regulated by planning procedures.

1.4 Types and sources of land disturbance and dereliction

1.4.1 Routes to degradation and back to self-sustainability

There are two main categories of regulated (in the sense that they are legislated for) use that lead to a degradation in the ability of land to support a self-sustaining ecosystem. These are *temporary* and *permanent* uses (Table 1.4). These definitions are anthropocentric, as in the longest term all of these uses will be ended as crustal material is recycled, and eventually the planet itself will be recycled, but this is beyond the scope of this book, and so these terms have utility for our purposes. There is a third category, *ecosystem depletion*, which results from unregulated activity in otherwise undisturbed areas, such as atmospheric pollution or poaching wild animals, which can also lead to degradation and to dysfunctional ecosystems. These three modes of land-use are interlinked by changes in entropy and biodiversity as shown in Fig. 1.6.

Temporary uses These are uses where there is a planned programme of use, perhaps to extract a resource or act as storage for other projects, followed by a prescribed reclamation or restoration programme. In these cases, the financial dimension of the after-use period has been included in the plan and is therefore most likely to be successful in its outcome, within our current limits of technology and understanding.

1. **Quarries**. Quarries tend to operate over a long timescale. Often the minerals involved are of low value, and there is insufficient overburden material to make good the original contours. Limestone quarries are a good example, and there are several examples of old quarries being put to imaginative end-uses, such as recreational water bodies, and conservation end-uses by means of 'restoration blasting'. The latter technique

Table 1.4 Relationship between land-uses and mechanisms of change

Temporary	Strip or opencast mining
	Quarrying
	Civil engineering
	Landfill
Permanent	Agro-froestry
	Civil engineering
	Built structures
	Structured amenity and recreation
Ecosystem depletion	Uncontrolled recreation
	Wildlife hunting or poaching
	Theft of genetic resources
	Pollution from off-site

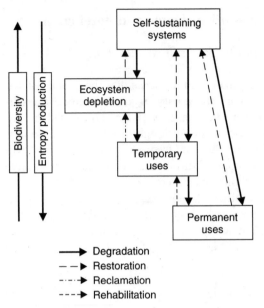

Fig. 1.6 Relationships between land-uses, biodiversity and entropy production.

involves recreating geomorphological features by controlling the last production blast.

2. **Strip or open-cast mines**. Open-cast mining is really another form of quarrying, but in many cases carried out over a shorter time period. Large areas of land are disrupted each year by strip or open-cast mining. The principal mineral excavated in this way is coal, but many others such as bauxite and clay are mined in this fashion. Essentially a large area of soil is removed, followed by the subsoil and overburden, exposing the mineral layers. This type of mining can be very cost-effective, and practised at great depth dependent upon the value of the mineral reserve. In some cases there is an added benefit of dealing with surface degradation at the same time.

3. **Civil engineering**. There are many civil engineering projects that result in permanent structures but involve temporary activities; for example, the need to store materials or to gain access to geological strata. Tunnelling and road building would fit into this category, particularly where cut and fill methods are used.

4. **Landfill** sites are large voids in the ground which have been used to dispose of domestic and industrial wastes. Landfills are associated with problems such as little or no cover materials, toxic leachate, and generation of landfill gas. Sites used for landfill may be restored to both soft and hard end-uses, and require a high engineering input to be successful.

Permanent uses   These are uses where there is no long-term intention to return the land back to self-sustainability. As a consequence of this, there is no in-built *a priori* economic mechanism for carrying out a reclamation or restoration programme. The method and end-point of reclamation is determined by the requirements of regulations and cost. When the difference between the purchase cost of the degraded site and its sale price in a useable form exceeds that of the treatment programme then restoration or reclamation is possible. If regulations require that the site be treated to a particular standard or incorporate particular features (including sustainability) then the cost of this will be reflected in the purchase price. Facilitation of reclamation to particular standards or end-uses may be assisted by government grants.

1. **Agro-forestry**. This is by far the major use resulting in the disturbance of naturally occurring ecosystems. Many landscapes, such as the cereal area of East Anglia and the commercial forestry plantations of Scotland, are ecological deserts, populated by one or two species of plant. Major challenges to the process of ecological restoration are to restore these areas to a sustainable land-use whilst maintaining their economic viability.
2. **Civil engineering**. Reclamation, at present, is principally a civil engineering activity. In addition, many civil engineering operations require some degree of restoration or reclamation. Examples are highway construction, pipeline construction and sea defence work.
3. **Built structures**. This category, which includes those uses such as factories, housing and office accommodation, results in land being unavailable to contribute to biogeochemical processes. The activities on site may also cause pollution. *Direct contamination* occurs when there is an industry producing a highly refined product, or complex waste, which may be released accidentally or otherwise on-site, or sometimes off-site. Smelters and gasworks are good examples here.
4. **Economic decline**. In many areas, general decline in industrial and commercial activity leads to the demise of whole communities. This can lead to large areas of land with a mixture of uses requiring redevelopment or restoration. In such areas land of all types can be further degraded by fly-tipping resulting from a general disregard for the environment.

Ecosystem depletion   In many areas of the world ecosystem depletion is the major threat to the self-sustaining function of ecosystems (Dooge *et al.*, 1992). Such depletion takes many forms, some subtle, some obvious.

1. **Uncontrolled recreation**. Many a pleasing, gently undulating landscape has been transformed into unstable mud runs by uncontrolled use by cross-country motorcycles, and damage to upland vegetation as on the Pennine Way. There is either no legislation to prevent such activities taking place or there are insufficient resources set aside to enforce it. In either case, there will therefore be inadequate resources available to repair damaged areas, except where local populations become vociferous.

2. **Wildlife hunting or poaching**. There are many cases where well-controlled game-hunting activities form a well-regulated and efficient part of a harmonious landscape. Unfortunately, all too often there is unregulated and unsustainable pressure on certain conspicuous or otherwise valuable species. The species may become extinct, and its role in the efficient recycling of matter and energy within an ecosystem lost.

3. **Theft of genetic resources**. There are many examples where wholesale removal of economically valuable resources, such as certain tree species, occurs to the great detriment of the ecosystem and its function. In many instances there is no biological potential left, as all members and propagules have been removed, or there is no in-built redundancy in the system able to take over that species' function.

4. **Pollution from off-site**. The principal source of widespread contamination of the atmosphere is as a result of the combustion of fossil fuel stocks, notably coal, for energy production. This activity has resulted in the deposition of large amounts of sulphur and nitrogen in previously pristine sites, and is commonly known as 'acid rain'. Metal ore smelters can give rise to localized increases in atmospheric pollutant levels and there have been noteworthy cases of contamination by nuclear power and processing plants, and liquids and gases may migrate off-site.

1.4.2 Pollution

One of the major consequences of both types of controlled land-use (temporary and permanent) is the generation of pollutants. Definition of pollution would appear to be straightforward at first but, in reality, it is a complex phenomenon. For our purposes any phenomenon which takes a characteristic of a site outside its normal limits of variability can be said to be polluting, providing that it was not polluted in the first place. This catches most instances and covers everything from light and thermal pollution through to dumping of radioactive waste.

Pollution can arise from both point sources and background dispersion. Point sources include *in situ* contamination where, for example, contamination arises on the site of an industrial process, either through spillage or deliberate waste disposal on sites. There may also be accidental spills in transit, or atmospheric pollution from a plume blowing from a smelter with the prevailing wind. Migration of contaminants into natural soils and groundwater may occur both on- and off-site. The pollution of the atmosphere is widespread and has significant effects on all types of land uses (Wellburn, 1994).

Polluting industries   The following is a list of industries commonly associated with contaminated land (Smith, 1985):

1. Gasworks and similar sites.

2. Scrapyards.
3. Railway land.
4. Sewage works and farms.
5. Mining and extractive industries.
6. Waste disposal.
7. Metal smelting and refining.
8. Metal treatment and finishing.
9. Paint and graphics.
10. Pharmaceutical, perfume, cosmetics and toiletries.
11. Pesticide manufacture or use.
12. Iron or steel works.
13. Tanning and associated trades.
14. Wood preserving.
15. Dockyards.
16. Chemicals.
17. Acid/alkali plants.
18. Oil production, refining and storage.
19. Explosives industry.
20. Asbestos manufacture and use.

Such uses can give rise to a wide range of contaminants, from simple inorganic contaminants such as heavy metals to complex organics such as polychlorinated biphenyls (PCBs), which have a wide range of persistence and treatability.

Naturally poor soils   Some soils are naturally poor in nutrients or water relations, or may be naturally contaminated. The decision to alter them for a productive use, biological or engineered, is essentially a decision to *develop* as a perfectly self-sustaining ecosystem may be disrupted in the process.

1.5   Derelict land survey

The principal sources of derelict and contaminated land are industry and mining, or the fall-out from these activities. Military dereliction is also an important source which has become more prominent in the late twentieth century because of the decline in military forces and bases after the end of the 'cold war'. Derelict land in England has been surveyed in 1974, 1982, 1988 and 1994. The Scottish Office and Welsh Development Agency have also carried out surveys within Scotland and Wales. These surveys, which are the most comprehensive nationwide surveys in Europe, provide a clear picture of the sources and types of derelict land in an industrialized country. The results from the 1988 survey are summarized in Table 1.5.

The survey recorded 40 500 ha of derelict land (1 April 1988), 78 per cent of which was considered to justify reclamation. This was a fall of 11 per cent in

Table 1.5   The amount of derelict land in England and the area justifying reclamation by type of dereliction, 1 April 1988 (DoE, 1991)

Source of derelict land	Area		Area justifying reclamation		% justifying reclamation
	ha	%	ha	%	
Spoil heaps	11 900	29	7 500	24	63
Colliery spoil heaps	4 700	12	4 400	14	93
Metalliferous spoil heaps	4 800	12	1 300	4	27
Other spoil heaps	2 400	6	1 800	6	75
Excavations and pits	6 000	15	4 400	14	73
Military, etc., dereliction	2 600	6	2 100	7	80
Derelict railway land	6 400	16	5 000	16	79
Mining subsidence, etc.	1 000	3	900	3	90
General industrial dereliction	8 500	21	8 000	25	94
Other forms of dereliction	4 100	10	3 800	12	92
Total	40 500	100%	31 600	100%	78%

Note: Percentages are rounded to the nearest integer and areas to the nearest 100 hectares. Component figures may not sum to the independently rounded totals.

total derelict land since 1982 (45 700 ha), compared with an increase of 6 per cent between 1974 (43 300 ha) and 1982. (Compare with USA 1 784 517 ha derelict land in 1974.) The survey shows that spoil heaps accounting for 29 per cent were the most common form of derelict land and, of these, 80 per cent were accounted for by colliery spoil heaps and metalliferous mine spoil heaps in almost equal proportions. General industrial dereliction accounting for 21 per cent of derelict land was the next most extensive type. Interestingly, the local authorities making the returns considered that only a relatively small proportion of metalliferous spoil heaps justified reclamation, whereas they considered 93 per cent of colliery spoil heaps justified reclamation. Part of the reason for this may be that metalliferous spoil heaps are often in more remote rural areas and may have become naturally vegetated, with the result that it was considered that there was less justification for their reclamation.

Between 1982 and 1988 some 14 000 ha of derelict land had been reclaimed, 61 per cent by local authorities and 90 per cent of this with the aid of the Derelict Land Grant. Of the land reclaimed, 27 per cent was for hard end-use (building of various sorts), and 33 per cent for public open space (including sports and other amenities). Public sector reclaimed 20 per cent to hard end-use, private sector 40 per cent. Derelict land was almost equally divided between urban and rural areas. The amounts of land derelict and justifying reclamation fluctuated between the surveys but remained at approximately 0.25 per cent of the land surface of England (Table 1.6).

There are no national surveys of contaminated land in the United Kingdom and the derelict land survey conducted by the Department of the Environment does not require contaminated land to be identified. One of the difficulties in establishing the amount of contaminated land is that using the generally accepted definition of contaminated land requires knowledge not only that substances are present at a site but also that these substances will, or are likely

Table 1.6 Trends in the amount of derelict land in the United Kingdom (DoE, 1991)

Year of survey	Area of derelict land (ha)	Area of derelict land justifying reclamation (ha)
1974	43 300	33 100
1982	45 700	34 300
1988	40 500	31 600

to, cause damage. To establish such information requires site investigation. Appropriate site investigations have not been carried out for many sites and such information has not been centrally collected. Surveys of 'potentially contaminated land', based on the historic use of sites, have been carried out in the United Kingdom. These surveys rely on information on the previous use of sites to determine whether they are likely to be contaminated. The earliest such survey is perhaps one carried out in Wales in 1982 and subsequently updated (Welsh Office, 1988) which identified over 700 sites greater than 0.5 ha in area which were potentially contaminated. Later surveys were carried out at a local authority level in preparation for or in response to proposals by the United Kingdom government to require all local authorities to prepare registers of land which may be contaminated. These surveys indicated that, based on previous use, many local authority areas contained thousands of sites which could be termed 'potentially contaminated'. The importance of previous use in determining the contamination status of a site will be returned to in Chapter 4.

An indication of the effects of different types of dereliction/contamination is shown in Fig. 1.7.

1.6 Legislation and regulation governing derelict/contaminated land

Legislation and regulation in the United Kingdom governing derelict and contaminated land was subject to considerable change between 1989 and 1995. Motivation for this change was in response to both European Commission directives and a greater popular environmental awareness influencing politicians and governments. Principal legislation was the Environmental Protection Act 1990 and the Water Resources Act 1991, both of which brought together a number of previous acts and incorporated new legislation. A summary of the principal provisions relating to derelict and contaminated land is given in Table 1.7.

Much reclamation is also carried out as part of active mining, and here reclamation is governed by mineral planning, and environmental assessment legislation and guidance. The exploitation of minerals in the United Kingdom is generally not allowed to proceed without planning conditions which deter-

Former uses	Poor natural ground conditions	Variable ground conditions	Settlement of filled ground	Underground obstructions, buried pits/tanks	Underground voids	Bodies of open water	Health hazards	Chemical contamination	Aggressive ground conditions	Combustible materials	Expansive materials	Explosive materials	Gas emission	Leachate production	Steep slopes
Green field site	●	●													
Residential	●	●	●	●	●										
Chemical works	●	●	●	●	●	●	●	●	●	●	●		●	●	
Steelworks	●	●	●	●	●	●	●	●	●	●			●	●	
Paper/printing works	●	●		●			●	●	●	●			●	●	
Munitions production	●	●		●	●		●	●	●	●		●		●	
Oil refinery	●	●		●			●	●	●	●	●	●	●		
Gasworks	●	●	●	●			●	●	●	●			●	●	
Sewage works	●	●	●	●	●	●	●	●	●				●	●	
Abattoir	●	●		●			●								
Tannery	●	●		●			●	●	●				●	●	
Scrapyard	●	●	●	●	●		●	●	●			●		●	
Mineral extraction	●	●	●	●	●	●	●	●	●	●		●		●	●
Metal mining/smelting	●	●		●	●	●	●	●	●					●	
Railway sidings	●	●		●		●	●	●	●			●		●	
Roads/airports	●	●		●	●		●		●			●			
Docks/canals	●	●	●	●		●	●					●	●		
Spoil heaps/slag heaps	●	●	●				●	●	●	●				●	●
Industrial landfill	●	●	●				●	●	●	●		●	●	●	
Domestic landfill	●	●	●				●	●	●				●	●	
Fly-tipping	●	●	●				●	●	●	●		●	●	●	●
Categories at risk															
Site workers				●	●		●	●	●	●	●		●	●	●
Residents nearby						●	●	●					●	●	●
Residents on-site						●	●	●	●				●	●	●
Structural stability	●	●	●	●	●			●	●	●		●	●		●
Building materials								●	●	●			●	●	
Underground services	●	●	●					●	●	●			●		
Domestic gardens						●	●	●	●				●	●	
Vegetation							●	●	●				●	●	●
Water resources						●	●	●	●				●	●	

Fig. 1.7   Characteristics of derelict/contaminated sites.

mine the extent and nature of reclamation to be carried out subsequent to the minerals being won. Many mineral operations carry out restoration as mining progresses with substantial amounts of land being returned to beneficial use whilst the mine is still operational. A substantial amount of guidance on the investigation and reclamation of derelict and contaminated land and land in active mining has been issued by the United Kingdom government. This guidance is referred to in subsequent chapters of this book.

Table 1.7 A summary of United Kingdom legislation and regulation relevant to derelict and contaminated land

Legislation	Principal provisions relating to contaminated/derelict land
Water Resources Act 1991	Incorporates Water Act 1989, which established the National Rivers Authority, and other water-related legislation. Provides for the establishment of water protection of 'controlled waters' which are: relevant territorial waters; coastal waters; inland freshwaters and groundwaters. It is an offence to 'cause or knowingly permit any poisonous, noxious or polluting matter or any solid waste matter to enter any controlled water'. This provision would include the pollution of surface or groundwater arising from contaminated land.
Environmental Protection Act 1990 Environment Act 1995	Incorporates and consolidates previous environmental legislation. Contains wide-ranging new provisions for the regulation and control of waste disposal. Provides for the establishment of a new Environmental Agency incorporating the responsibilities of the National Rivers Authority, Her Majesty's Inspectorate of Pollution and the local authority waste regulation function. Original provision (S143) for compulsory registers of potentially contaminated land dropped by the Government in 1994. Provision for voluntary registers introduced in 1995 Act.
Derelict Land Act 1982	Provides for grant aid for derelict land reclamation in England and Wales. Scope of schemes for which grant aid may be given is wide and includes contaminated land, providing it is derelict.

1.6.1   United Kingdom

The first legislation to be passed was the *Town and Country Planning Act 1944*, enabling local authorities to acquire derelict land in order to bring it back into use. This was little used and was repealed, to be replaced by the *Town and Country Planning Act 1947*. Following this, there was Section 89 of the *National Parks and Access to the Countryside Act 1949* (amended by Section 6 of the *Local Authorities (Land) Act 1963*), which allows county and district councils to carry out works to bring derelict land back into use, and to improve the appearance of generally neglected land. Land can also be acquired under the *Town and Country Planning Act 1971* for development of residential or amenity space, which can include derelict land.

The *Mines and Quarries (Tips) Act 1969* placed the onus of responsibility on local authorities for ensuring the stability of spoil tips, which usually includes restoration work.

Much of this legislation makes grants available to the local authorities to enable the work to be done. This can vary from 50 to 75 per cent of the total cost of the works, depending on the category of the land, and the economic status of the area. In some cases 100 per cent grants are available.

The working of the Opencast Executive of British Coal has always been tightly controlled by the *Mines and Quarries Act*, directing the Executive to draw up detailed restoration plans prior to permission being granted for work to begin.

The *Minerals Working Act 1951* provided a fund for restoration of ironstone workings.

In the mid-1970s some significant pieces of legislation were passed. The *Town and Country Planning General Development Order 1973* gave planning authorities the power to require mineral extractors to provide for adequate disposal and reclamation of spoil materials. The *Control of Pollution Act 1974* provided for the systematic collection and disposal of household, industrial and commercial wastes, but not those arising from mineral extraction.

There has been extensive use by local authorities of *Derelict Land Grants* from the Department of the Environment to reclaim and restore derelict land.

The *Derelict Land Act 1982* Section 3 substituted the Parks Act of 1949, and conferred specific powers on local authorities.

1.6.2   United States

Until 1977 individual states exercised control over restoration of mined land. The *Federal Surface Mining Act 1977* created overall regulations for reclamation and restoration on coal-mined land. This introduced minimum standards with regards to replacement of contours, segregation and replacement of topsoil (something British Coal had been doing on open-cast sites since the mid-1950s), vegetation establishment and hydrological protection. A bond is paid by the mining company and is not repaid until the land is satisfactorily restored. There is also a levy on coal production to allow previously unrestored land to be reclaimed and to alleviate off-site impacts.

1.6.3   Germany

The former West Germany also has tight control over mining reclamation with detailed consultation and planning in the Ruhr Valley overseen by the *Siedlungsverband Ruhrkohlenbezirk*.

1.6.4   Central Europe

The needs of industrial production have been paramount in the former Soviet bloc countries. As a result restoration was often done as an afterthought at best, with poorly resourced reclamation authorities given sites after the mining authorities have exhausted the land, very often with no regard to vegetation, contours or hydrology. This is one of the primary legacies of a state-controlled system bent on modernization. Nevertheless considerable knowledge of the use of soil materials in restoration has been built up in some of these countries and some good reclamation schemes to agriculture carried out.

1.7 Conclusions

It is clear that there are many routes by which ecosystems may become degraded and therefore be in need of repair. Much damage results from inefficient use or overuse of resources, lack of understanding of ecosystem function but sometimes by wilful negligence. There is a sequence of events which should, ideally, be followed when a change in land-use is planned or occurs as a result of external factors (Fig. 1.8). The chapters ahead set out how we may go about the business of assessing the damage and putting it right, following this sequence.

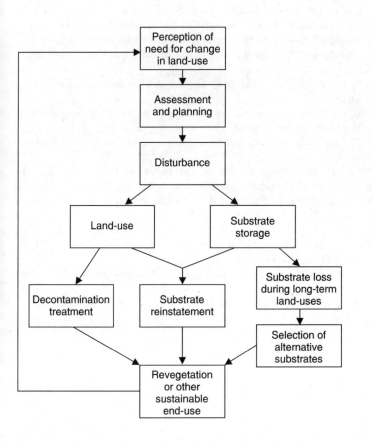

Fig. 1.8   Ideal sequence of events during change of land-use.

References

BRADSHAW, A.D. and CHADWICK, M.J. (1980). *The restoration of land: the ecology and reclamation of derelict and degraded land*. Blackwell, Oxford.

DEANGELIS, D.L. (1992). *Dynamics of nutrient cycling and foodwebs*. Chapman and Hall, London.

DEPARTMENT OF THE ENVIRONMENT (1991). *Survey of derelict land in England 1988*. HMSO, London.

DOOGE, J.C.I., GOODMAN, G.T., LA RIVIERE, J.W.M., MARTON LEFEVRE, J., O'RIORDAN, T., PRADERIE, F. and BRENNAN, M. (eds) (1992). *An agenda for science for environment and development in the 21st century*. Cambridge University Press, Cambridge.

EUROPEAN COMMISSION (1994). *Competitiveness and cohesion: trends in the regions*. Fifth periodic report on the social and economic situation and development of the regions of the community. European Commission, Luxembourg.

GRIME, J.P. (1979). *Plant strategies and vegetation processes*. John Wiley and Sons, Chichester.

GUNN, A.S. (1991). The restoration of species and natural environments. *Environ. Ethics*, **13**, 291–310.

HARRIS, J.A. and BIRCH, P. (1992). Land reclamation and restoration. In Fry, J.C., Gadd, G.M., Herbert, R.A., Jones, C.W. and Watson-Craik, I.A. (eds) *Microbial control of pollution*, SGM Symposium volume 48. Cambridge University Press, Cambridge.

LOCKWOOD, J.L. and PIMM, S.L. (1994). Species: would any of them be missed? *Curr. Biol.*, **4**(5), 455–457.

MCNEELY, J. A. (1994). Lessons from the past: forests and biodiversity. *Biodiversity Conserv.*, **3**, 3–20.

MEADOWS, D.H., MEADOWS, D.L. and RANDERS, J. (1992). *Beyond the limits: global collapse or a sustainable future?* Earthscan, London.

ODUM, E.P. (1993). *Ecology and our endangered life-support systems*, 2nd Edition. Sinauer, Sunderland, Massachusetts.

O'RIORDAN, T. (1995). Environmental science on the move. In O'Riordan, T. (ed.) *Environmental science for environmental management*. Longman, Harlow.

SMITH, M. A. (1985). Background. In Smith, M.A. (ed.) *Contaminated land: reclamation and treatment*. Plenum Press, New York.

STOLZ, J.F., BOTKIN, D.B. and DASTOOR, M.N. (1989). The integral biosphere. In Rambler, M.B., Margulis, L. and Fester, R. (eds) *Global ecology: towards a science of the biosphere*. Academic Press, San Diego.

WELLBURN, A. (1994). *Air pollution and climate change: the biological impact*. Longman, Harlow.

WELSH OFFICE (1988). *Survey of contaminated land in Wales*. Welsh Office, Cardiff.

WILSON, E.O. (1992). *The diversity of life*. Belknapp Harvard, Cambridge, Massachusetts.

Further reading

O'RIORDAN, T. (ed.) (1995). *Environmental science for environmental management*. Longman, Harlow.

RICKLEFS, R.E. (1993). *The economy of nature*. W. H. Freeman, New York.

Chapter 2

Assessment strategies

2.1 Defining what constitutes environmental quality

The basic question that must be asked when designing a management programme for land is 'will it result in a sustainable end-use?' It is important to discriminate between self-sustaining uses, i.e. those with integrated production, consumption and decomposition functions, and those that will always require a level of intervention by man, such as agro-forestry or building construction. Without proper maintenance buildings will decay, and to prevent decay maintenance procedures need to be incorporated in the management plan for a site. At the construction stage there are usually more immediate problems to be addressed such as 'is the site geotechnically stable – if not, how is this to be remedied?' or 'what level of contamination is there – can it be treated *in situ* or will it need to be removed and replaced?' There will need to be a sophisticated integration of biological, geotechnical, hydrological and economic factors. If any one of these is missing the system is likely to fail.

One definition of quality is that of the flexibility of land as a resource, defined by O'Riordan (1971) in the following way: 'a resource is an attribute of the environment appraised by man to be of value over time within constraints imposed by the social, political and institutional framework'. This is broad enough to encompass any definition of 'value' – and is the real source of where debate over land-use and management lies, not in the fine detail of assessment methodology. O'Riordan (1971) has indicated that:

1. Land resources are finite.
2. Cheap technological 'fixes' are not always available.
3. Development of different resources are not independent of one another.

To manage land resources responsibly, they should be used in ways that maintain their future *flexibility*, i.e. maximize the number of options available once the resource is required for other uses. Options should include returning the land to the 'natural' state after use. It is because a return to self-sustaining systems was not considered in the past that land reclamation and restoration is now a major human activity. We must not forget, however, that the original developments were carried out with the prevailing knowledge and economic conditions. Without the economic activity of the time, we may not have arrived at our present state of knowledge. It is both easy and dangerous to condemn with the benefit of hindsight. The major cause of this degradation has not been in fact that the activity was carried out, but that little or no provision has been made for the maintenance of ecological diversity *in terms of the economics of the site*, even if the maintenance of that diversity has to be carried out off-site. If this had been done then many proposed developments would have been deemed uneconomic or even impossible to achieve – we may then decide to 'foot the bill' as a society, or to take the decision that the development is unacceptable. It is usually at this point that decisions of a political nature must be made.

We will take the stance, initially, of the ideal approach to assessment, i.e. prior to any change in land-use from natural or agro-forestry, the latter being so ingrained and widespread in human society that it would be all but impossible to disentangle it from the more conservative landscape uses, although in Chapter 4 full consideration will be given to the needs of sites subject to industrial use as the starting point.

2.2   Assessment of resources – overview

A number of issues need to be addressed when considering the potential uses of land, including those of economic returns, social pressures, conservation of potential uses after the proposed development has finished and maximization of agricultural productivity; in other words an assessment of its environmental impact. Formalized environmental impact assessment (EIA) will be considered later in this chapter. Firstly we will consider how procedures based on single disciplines have proceeded, and we will attempt to point to how future assessment will be achieved.

Clearly, when proposing a change in use, reclamation and restoration must be included in the planning phase and *proper economic provision be made for its successful completion*. As outlined above, without this any reclamation or restoration scheme is unlikely to be successful.

2.2.1   The question of scale

The first issue to be tackled when assessing a potential change in land-use is 'at what scale is the potential effect of the change of use to be considered?' This has two dimensions, those of size and time. Usually considerations are taken at the level of the site; substrate geotechnical condition and quality, existing built infrastructure and hydrogeology. Regulations at national and European Union (EU) level and considerations of liability have led to an extension of site level factors to include considerations of impacts arriving from off-site, and substances generated by the land-use and likely to be transmitted off-site. There is increasing recognition that a wider view should be taken such that if the resource found on this site is lost as a result of the change of use, will it be replaceable on a larger scale? An example might be where endangered species are found on only a few sites, and the change in use would lead to an irreversible loss of biological potential.

Quite different land-uses are ultimately interconnected at some scale, as illustrated in Fig. 2.1. Through this approach it is possible to demonstrate the need to consider land-use changes at larger scales than has been practised in the past, including how restoration or reclamation schemes serve to benefit the whole landscape, in an integrated way. Klijn and de Haes (1994) have proposed a series of levels at which parameters may be considered (Table 2.1), based on three considerations:

1. Relation to the characteristic classified, based on an integrated or 'holistic' approach.
2. Relation to the most commonly used nomenclature on a global scale.
3. Easy translation into all European languages.

This indicates the need to integrate different disciplines in assessment, and to give full weight to the interaction between different ecosystem parameters. What is most important is that this includes human uses, such as the built environment. In practice this occurs at present when things are considered at local, regional and national scale. In certain cases, where a species or geological feature is found only in one locality, this should be extended to continental and global scales.

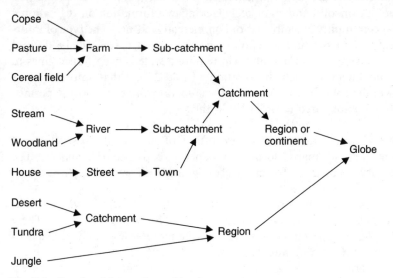

Fig. 2.1 Levels of integration of landscapes.

Table 2.1 A possible hierarchical classification for examining land-use (Klijn and de Haes, 1994)

Class	Description
Ecozones	Corresponds largely to global climate zones, such as the Arctic, temperate zone and tropical zones
Ecoprovinces	Corresponding to large-scale geological and geomorphological characteristics, such as the Rocky Mountains of the USA or the Great Glen of Scotland
Ecoregions	A more detailed division of the ecoprovince based on the differences in solid geology
Ecodistricts	These relate to features which may slowly change with time, such as soil series related to bedrock, and the hydrologic function
Ecosections	Corresponding to individual geomorphological features such as valleys and slopes
Ecoseries	This corresponds to the abiotic factors of the substrate with respect to plant growth, including soil, ground and surface waters. Different vegetational assemblages may be found on a single ecoseries due to different successional stages
Ecotopes	Generally referring to a defined assemblage such as deciduous forest, peat bog or agro-forestry unit, e.g. a field
Eco-element	Individual, but coherent feature within an ecotope. This could be a patch of dominants within a woodland, or a hedgerow in a field

We will now consider some of the approaches taken to assess impacts on sites and land capability, including complex and integrated approaches such as environmental impact assessment and catchment function surveys.

2.2.2 Types of information

The organization of information on the status of particular areas is extremely patchy. Some areas are well characterized for a particular set of properties, such as geology or soils, and completely lacking in information on vegetational and animal community assemblages or commercial land-use. The type of information required to produce a definitive assessment of a site will be derived from a wide variety of sources, and will require a wide range of disciplines to interpret and integrate such information (Table 2.2). What information is required will depend on what is going to happen to the site, and the permissions that will be required in order to do this.

Site history Local libraries and record offices often offer the best source of information for old maps, local newspapers, road and site names. The Ordnance Survey has maps for many areas going back several centuries.

Current use An assessment of the current use of the site includes visual 'walk-around' surveys where site access is not possible, for whatever reason. There will also be planning permissions, which may be more or less detailed, depending upon when they were granted. Statutory instruments of control may exist, which are available to the local authority to enable it to enter premises, wherein those processes are listed. The National Rivers Authority issues discharge

Table 2.2   Types of information

Type of information	Source
History	Maps
	Newspapers
	Photographs
	Place and road names
	Books
	Past planning applications
	Court proceedings
	Process layouts
Current use	Visual survey
	Economic indicators
	Council records
	Business directories
	Discharge consents
Landform	Land surveying
	Maps
Biogeochemical status	Soil maps
	Geological maps
	Hydrogeological maps
	Groundwater vulnerability maps
	Biological monitoring
	Environmental sampling and analysis
	Public utility records

consents, where the type and concentrations of effluent allowed to leave the site are given.

Landform   At the simplest level, contour maps (e.g. Ordnance Survey maps) are available indicating the shape and elevation of the site. More sophisticated mapping may be required in some circumstances, which must be carried out by a land surveyor; this will also have implications for the hydrology of the site. The impact of roads and other structures on the site must also be taken into account at this stage. Visual assessment would be carried out by a landscape architect.

Biogeochemical status   The site under consideration may hardly function biogeochemically when an industrialized site is being considered, or may be fully functional if an undisturbed site is being considered. There is a wide range of techniques available for determining the structure, function and integration of the biological, geological and hydrological status of a site, which are given in more detail below.

Remote sensing   This is based on satellite technology, and increasingly higher resolutions are being achieved, showing smaller terrestrial features than has been possible in the past. There are very large programmes to monitor global environments, particularly as they relate to climate change. This type of programme is not restricted to visible wavelengths; indeed much useful information with regard to land-use and disturbance has come from using infrared wavelengths. This is due to the different reflectances of different vegetation types, and the warming of soil due to the action of micro-organisms, engendered by cultivation and other disturbances.

Aerial photography This can be particularly useful when a library of plates taken at different timescales is available, enabling progress of succession, disturbance, conservation and urbanization to be assessed.

2.2.3 Organization of information

Most of the assessments of site characteristics are still based, currently, on traditional, rigorous methods using field and laboratory equipment. Increasingly, however, more sophisticated techniques are being employed such as satellite imagery and aerial photography, outlined above. The major advances, however, have come in the development of extremely powerful computer-resident relational databases tied to graphical representations of sites, areas and regions. Data may be input and manipulated with great ease, and this is described below.

Geographical Information Systems (GIS) These are computer-resident programs containing virtual maps, upon which data may be layered at a number of levels. In the early days of GIS this allowed simple production density maps and many of these are now available as PC (personal computer) packages. More recently, as the power of computing has advanced, the systems allow far more sophisticated inter- and extrapolations to be made. Such questions as 'which areas have both pH of 6–6.5 and spruce plantations?' or 'which is the best land for planting alder?' may be asked. Further, planners can identify areas where a service is lacking or a particular land-use type is present. As time progresses, it will be possible to interrogate GIS on-site using hand-held sets fitted with global positioning equipment. Immediate 'ground-proof' data can be input, giving an instant analysis of the progress of restoration or degradation.

The functional units used in a GIS package are shown in Fig. 2.2. The data may be captured in a number of ways, by scanning existing maps, by digitizing them with a hand-held device on a digitizing table, or by direct input of files already existing from such sources as the Ordnance Survey, or the Soil Survey and Land Research Centre (formerly the Soil Survey of England and Wales). The type of information for each point or polygon may be as follows: soil class, textural class, slope, concentration of chemicals, horizonation (depth of soil profiles), organic matter content and rainfall.

This list could be extended to cover any aspect of the point which has a discrete value, including the presence of particular plant species, although it

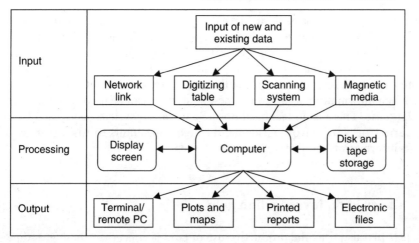

Fig. 2.2   Flow of information in a GIS.

must be recognized that such data have a far faster turnover time than the hydrogeologic factors. Also a series of these points can be grouped together and be given the label 'site' and this in turn can be given a set of values for site characteristics, such as the presence and type of vegetational assemblages. At this level the concept of ecotopes and ecoseries as suggested by Klijn and de Haes (1994) would be both pertinent and useful.

Data of this type have been available for some time, but only with the advent of modern computing power has the inspection, integration and easy manipulation of such data been possible. It is also important to note that there may be mismatches in data sets, particularly when time has elapsed between their collection and collation, that analytical techniques have changed, and that there may be simple misalignment occurring when maps of the same scale have been produced to different longitude/latitude projections.

Once the data have been input and correctly aligned it may be manipulated to produce the desired output, which may be in the form of maps of one type of characteristic, e.g. isobars or contours, or it could be instructed to produce a map showing the position and extent of sites sharing common features, e.g. all soils within the clay loam texture class with 1100–1200 mm (40–50 inches) of rainfall per annum growing arable crops.

One problem still outstanding is the question of resolution. Some of the information is only available on 5 km grid intersections, while in other areas there may be extremely detailed maps where sampling has occurred on 10 or even 5 m grids. In developing nations the information may be only available from satellite imagery. In any case once entered in such a system the information may be maintained and updated, and the resolution improved.

Currently GIS is mainly used at regional level for planning and strategy purposes and for specific-site studies, but as more powerful systems are developed then it becomes possible to develop their use at site level, particularly when integrated with global positioning systems (GPS), which can locate a current position by interrogation of a satellite positioning system. By interrogation of a GIS database it will be possible for data held for a site to be instantly recalled, and to recall how the site is impacted by its surrounding area. When information is gained in a new study, this can be used to update the GIS database. There is no technical obstacle to such a system being used by a wide variety of people, from residents, through consultants to town-planners, to gain information on a site of interest to them.

2.3   Site-specific investigations

It is clear that as far as the productive use of land is concerned there are some sophisticated systems available from which to develop targets of restoration or reclamation. These approaches do not, however, cater for the need to assess the value of natural systems. Current practice for the assessment of land resources prior to a change in use, be they pristine, in-use or degraded, is focused on the investigation of specific sites. A developer will be expected to make investigations only to add to information already in the hands of the local planning authority gained from many previous investigations. There are a number of standard techniques available for the analysis of materials.

2.3.1   Design of sampling programmes

There are a number of protocols developed specifically for assessing the degree of pollution in contaminated land, and they will be covered in more detail in Chapter 4. When collecting data for biological and chemical analysis a number of decisions must be made as to:

1.   **Frequency of sampling**. There are several factors which vary seasonally and diurnally, and these may have to be taken into account. There is little value in assessing the summer migrant bird population of a site in the middle of winter. An experienced ecologist will be aware of this.
2.   **Number of samples**. By taking some initial samples a variance will be obtained. This allows for the calculation of how many samples need to be taken to reduce error to within defined limits, usually 5–10 per cent of the mean, allowing statistical differentiation between sites (Spellerberg, 1991).
3.   **Appropriate sampling techniques**. These are particular to each discipline and within disciplines such as soil science. For example soil samples for microbiological analysis may need to be transported to the laboratory immediately, whereas certain chemical and physical parameters may be measured on samples which have been stored dry for some time.

4. **Random sampling**. For the majority of ecological sampling random sampling patterns are essential, particularly when the underlying division of properties of a visually similar site are unknown, i.e. although a ploughed field may look homogeneous there may be differences in soil characteristics which are not evident until samples are analysed.

2.3.2 Chemical, physical and microbiological analysis

Standard methods One of the major stumbling blocks to developing ecosystem assessments is the lack of agreement as to how environmental variables should be measured. One of the underlying causes of this is that scientists are constantly refining techniques, improving accuracy, speed and volume of replication, and developing novel measurements. Consequently it has been exceedingly difficult to apply a method which may be agreed upon by everybody involved. This is particularly the case when biological systems are being measured, which may require local knowledge and a high degree of skill. In the late 1960s and early 1970s an attempt was made to provide several 'cookbooks' of *standard methods* for measuring environmental variables through the work of the *International Biological Monitoring Programme*. Several extremely useful texts were produced, providing standardized approaches. These approaches have largely been ignored by the international biological community, particularly by those working in small groups or on restricted ecosystems who will assert that they have developed their own methods which are particularly efficacious in their situation; they are probably right. The rule still applies that if a method has been through the peer review process and has subsequently been published, then it is worthy of publication although it will probably be superseded with time. Also the methods were not subject to the rigours of testing in the land development market where standards and certification are essential.

There is, however, one international programme which is attempting to reconcile these disparate threads. This is the *International Organization for Standardization*, which is represented in the UK by the *British Standards Institute*. Specialist panels consider methods, modify them and publish an international standard method for assaying a particular parameter. Such work has been found to be particularly useful for physicochemical methods, but has proved somewhat problematical in developing standards for measuring biological standards. There are available a number of texts which provide standard methods for the analysis of environmental samples (Table 2.3).

Bioassays These involve the extraction of a sample from the environment under investigation and exposing a test organism or organisms to the sample in a raw or dilute form. The Microtox^R system is one example of this, which uses the photoluminescent bacterium *Photobacterium phosphoreum*. A liquid sample is mixed with a suspension of the organisms and a control blank, and

Table 2.3   Texts for environmental analysis

Type of analysis	Title
General	HAYNES, R. (1982). *Environmental science methods.* Chapman and Hall, London ALLEN, S.E. (1989). *Chemical analysis of ecological materials.* Blackwell Science, Oxford
Soil	MINISTRY OF AGRICULTURE FISHERIES AND FOOD (1987). *The analysis of agricultural materials,* 3rd Edition, Reference Book 427. HMSO, London
Water	UNITED STATES ENVIRONMENTAL PROTECTION AGENCY (1979). *Methods for chemical analysis of water and wastes.* Environmental Monitoring and Support Lab., Las Vegas, Nevada DEPARTMENT OF THE ENVIRONMENT (1978 *et seq.*). *Methods for the examination of waters and associated materials.* HMSO, London
Air	BRIMBLECOMBE, P. (1986). *Air composition and chemistry.* Cambridge University Press, Cambridge RICHARDSON, D.H.S. (1987). *Biological monitors of pollution.* Royal Irish Academy, Dublin WORLD HEALTH ORGANIZATION (1987). *Air quality guidelines for Europe.* WHO, Copenhagen

the light output determined. The drop in signal has been found to be directly proportional to a variety of toxic chemicals, including those which are normal constituents of pristine environments, but in low concentrations.

A more comprehensive range of organisms were used by Hund and Traunspurger (1994) including bacteria, daphnids, algae plants, earthworms, nematodes and fish. These workers found good agreement, in most cases, with known toxicities of contaminants *in vitro*. Microbial nitrification was found to be a rapid indicator of the presence of toxic compounds, and the aquatic systems were capable of a high degree of precision as to what concentrations were the point at which toxic effects began. Hund and Traunspurger concluded that using a battery of tests was the best approach in such situations. The cost of assessing sites other than those that are contaminated would be prohibitively expensive.

Biological indices   One approach which has been developed, to give at least some indication of the effects of activity on functioning ecosystems, is that of the environmental index involving audit of *indicator species, elements* or *activities*. There have been a number of efforts to assess the biological quality of systems, and wildlife and biotic indices have been put to widespread use.

Wildlife indices include those based on endangered species, 'troubled' species and the 'threat number' (Perring and Farrell, 1983). This latter was developed in the United Kingdom to assess the conservation status of endangered flora, based on the changes in numbers of particular species in the Red Data Books, within 10 km grids. There are also species distribution maps available, but their coverage is extremely limited taxonomically. Those species that have a high level of public interest, such as butterflies and birds, have many field records available, but other species are rarely recorded other than by professional scientists and natural historians, and it will always be impossible to record every species and know its status.

The description of *vegetational* assemblages has always been the preserve of the field botanist and recently a nationwide standard has been developed under the auspices of the Nature Conservancy Council, by a number of academics including those at Unit of Vegetation Science at Lancaster University. This is the *National Vegetation Classification* (NVC) system, and sites are classified into types based on community structure (Rodwell *et al.*, 1989). This will reflect the broad community type such as woodland or meadow, and will give details of the dominant species and, most importantly, a *class number*. This offers a real opportunity for integrating a true ecosystem structure measure into the assessment of area subject to disturbance, and further refines the potential for providing a defined end-point for the restoration programme. What is required is that the NVC, which is essentially an audit, is integrated into a system which takes other ecosystem characteristics into consideration; essentially becoming a measure of ecological soundness.

The Unit of Comparative Plant Ecology (UCPE) has produced a database of plant species related to their responses to stress and disturbance (Grime *et al.*, 1988). From this has been derived a computer program (Functional Interpretation of Botanical Surveys, FIBS) which, when a description of the plant community, in terms of species and abundance, is entered, produces an output indicating whether the community is stressed, disturbed or in a quasi-stable competitive state. Further to this, subsequent samplings may be taken and input to reveal whether there has been a change in community and in which direction – more or less stressed/disturbed or reaching stability. This has great potential for use on sites that are subject to change, whether managed, natural or accidental, and could be used to indicate how successful a restoration/reclamation management strategy is in affecting community structure.

In the United States a programme of research to develop an understanding of the mechanisms underlying the response of vegetation to air pollution and other environmental stress has been published under the title of 'The Response of Plants to Interacting Stresses' or ROPIS (Goldstein and Ferson, 1994). This programme included exposure responses, biochemical and physiological responses, whole plant development, material cycling, computer modelling of nutrient dynamics and plant adaptations. These measurements were taken at a number of sites across the United States and the plants were subjected *in situ* to a variety of stresses in open-top chambers. It was concluded that there was potential for using this type of analysis for determining stresses on almost any site once the baseline has been established.

Microbiological indices are proving to be extremely effective in indicating the extent to which soils have been damaged, and what effect management programmes have on repairing ecosystem damage. Bentham *et al.* (1992) have demonstrated that, using just three characteristics of the soil microbial community, discrimination between sites of different type and function may be achieved which is far superior to conventional measurements. These are based on the following characteristics:

1. **Size**. The amount of biomass in the system which may be determined directly in the form of carbon or indirectly by extraction of ubiquitous cellular components such as adenosine triphosphate (ATP).
2. **Activity**. A measure of the rate of turnover of materials within the system, and export/import of nutrients.
3. **Composition**. The degree of biodiversity within the system, which may be related to the relative proportions of types of metabolic groups or broad divisions, such as the percentage of the total biomass which is fungal.

Measurements are made in the laboratory on soil samples taken from the field. These three characteristics are then plotted against one another to give a three-dimensional ordination, as demonstrated by Bentham *et al.* (1992). The points at which the major ecosystem types are situated are clear, with grasslands, pioneer communities, woodlands and scrub systems falling into discrete clusters, with the systems in the process of restoration or reclamation lying outside of their 'target' clusters. With more data from other sites it has become clear that there is a distinct course of changes during succession, reflected in the changes in the microbial community (Fig. 2.3). These changes may be explained by the change in quality and quantity of organic material available to the soil microbial community during the course of succession. Early on there

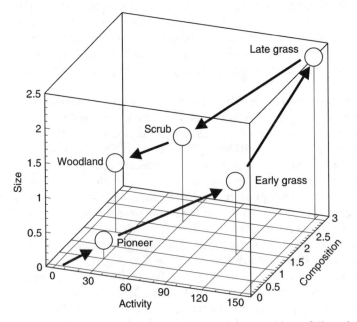

Fig. 2.3 Changes in the size, activity and composition of the microbial community during ecosystem succession.

is little material available as a result of exudation from the roots of pioneer species. Ruderal micro-organisms such as bacteria dominate the microbial biomass. As time and succession proceeds, more carbon substrates are released into the soil from the growing biomass of the primary producers, leading to greater microbial biomass. Eventually during the productive grassland phase, there is a very large biomass balanced between fungi and bacteria, and other microbial groups of less prominence, such as protozoa. Then, the size of the microbial biomass begins to decline as succession proceeds towards climax. This is not due to any less primary production, although there may now be some degree of stress in the system owing to limitations in mineral nutrients, but rather with the advent of *mycorrhizal* relationships there is less carbon being released into the bulk phase of the soil. As a result there is less carbon for the bulk phase biomass and, therefore, increasingly the biomass has to rely on the organic material arriving in the form of litter material. This material has a higher lignin content than is found in early and mid-successional plants and is therefore harder to break down. The decreased availability of the organic material leads to the second significant shift, from a community dominated by ruderal bacteria to one dominated by the stress-tolerant and competitive fungi.

This approach has allowed the progress of restoration programmes to be followed very closely, as well as indicating points of stability. It is intended that from this basis a thermodynamic model of these systems could be developed which would incorporate elements of the NVC model, i.e. incorporation of the three metabolic groups (primary producers, consumers and decomposers, plus detritivores if appropriate). This would not only allow prediction of effects of environmental perturbation on systems but also provide for a prescription for treatment, in terms of inputs of nutrients, organic matter, cultivation energy and genetic stocks.

The role of mycorrhizal relationships in stable systems is an important one and deserves some elaboration, as it will have a direct bearing on the success of restoration and reclamation programmes, as will be demonstrated later in Chapters 5 and 6. Mycorrhizae are the result of a relationship between the plant root and a fungus. The fungus can interact with the root physically in a number of ways depending upon the plant and fungal species, but essentially the plant provides the fungus with a supply of photosynthate (Peterson and Farquhar, 1994). The fungus is able to explore, by means of its fine filamentous mycelium, further than the root system, and into finer aggregates. This extension of the root system gives the plant access to a greater supply of phosphate and other minerals. Yet there is an even more important, and less obvious, consequence of this association. Firstly fungal hyphae of different species are able to fuse, linking themselves and therefore the plant partners, *also of different species*, together. This results in the formation of a vast network in forest systems where nutrients are efficiently conserved by the system as a whole. This has been demonstrated by Read and co-workers in radiolabelling experiments (Read, 1992). This extends beyond phosphate to nitrogen and carbon, and may include all elements transportable in this manner. When a tree dies, many of the nutrients it contains may be able to be transferred to the survivors in the stand, as decomposition by micro-organisms in the bulk phase alone will lead to large losses through respiration, excretion, leaching and run-off. The re-establishment of the plant–fungus network should be one of the principal goals of any restoration programme, as it is one of the keys to self-sustainability.

Water quality indicators   In many respects water bodies play a central role in the functioning of terrestrial ecosystems. On the catchment level they are extremely useful as 'integrators' of the changes that are occurring on the land that feeds water into them. Where changes occur in one part of the catchment as a result in changing from say agriculture to construction, there will be a concomitant change in the quality and quantity of the water draining from that land. Drainage water will be integrated downstream to give an indication of the overall impact of that change in land-use on the environmental quality of the catchment.

Biotic indices are based on the differential susceptibility of species within the community under study. The use of biotic indices has found particular, and very successful, application in the assessment of the pollution status of rivers and waterbodies. A number of indices have been developed, in particular the Neville Williams, the Biological Monitoring Working Party score (BMWP), the saprobian index, the Trent biotic index and the Chandler index. These indices take a long time to develop and are labour intensive in development, but are very efficient once established.

In recent years the biotic indicator approach has been refined and has become more sophisticated as demonstrated in the RIVPACS (RIver inVertebrate Prediction And Classification System) of the Institute of Freshwater Ecology (Wright *et al.*, 1993). This computer-resident program operates in two modes. Based on the input of a few simple physical and chemical data, it predicts what invertebrate community should be present at a particular site. The predicted community can then be compared with the actual community sampled. RIVPACS is an ideal tool for examining the effect of land-use change off-site, and enables the close monitoring of the effects of management on aquatic communities. The use of RIVPACS not only facilitates provision of a target but allows changes in management to ensure as rapid as possible achievement of targets.

These methods outlined above are currently largely in the domain of academic investigation. Increasingly, however, as defined communities and restoration standards are developed, it will be possible to apply them rigorously to assess the success or otherwise of restoration programmes. This may involve the application of just one type of assessment, such as the NVC, but ultimately it will be possible to integrate all these measures, giving meta-information, which only arises as a result of consideration of a multiplicity of data sets; e.g. how does a change in the vegetational assemblages on a site affect water quality? This can then form the basis of a predictive model of the interactions between different landscape elements such that it will be possible to plan change with a greater confidence in the ability to predict the effects such changes will bring about.

2.3.3 Geotechnical and hydrological factors

These factors have to be assessed in the majority of cases where there is likely to be an impact on soil water, or major ground works are proposed, such as road construction or quarrying. The characteristics to be investigated should include:

1. Extent and nature of materials underlying soils.
2. Load-bearing capacity, and ability to resist shear forces.
3. Location of ground water, and points of emergence.
4. Shedding point delineating catchments and sub-catchments.
5. Presence of old mine workings.

When the change in land-use involving restoration is limited as to area, this type of investigation may be unnecessary. It must be noted, however, that where a supply of water is essential, such as in wetland restoration, then the hydrology of the site becomes of paramount importance.

Krabbenhoft *et al.* (1993) have suggested a scheme for classifying reclaimed mine lands based on the definition and characterization of 'topoedaphic units'. This includes measurement of physical (top and subsoil depth, texture and water-holding capacity), chemical (sodium adsorption ratio, electrical conductivity, pH, concentrations of Ca, Na and Mg) and topographic (aspect, slope, position) characteristics. They concluded that the investigation of sites based on dividing areas into similar units using this system was superior to more traditional methods where the site is considered as a whole. It is interesting to note that this is almost exactly the process carried out in the United Kingdom when sites are to be disturbed for open-cast coal-mining purposes.

2.4 Land classification approaches

There are a number of land classification schemes organized for economic and agricultural purposes. These systems tend to focus on the ability of the land to support agro-forestry crops, which are distinct from those (although not exclusive from) systems designed to encompass all types of land-use (see sections 2.4.3 and 2.4.4). The schemes in the United Kingdom, United States and Canada share a common lineage, and may be compared directly. They are based on considerations of the following factors: climate, nutrients, soil, wetness, erosion, pattern, vegetation and topography.

The emphasis placed on each of these varies from system to system (Table 2.4). In some cases certain factors are omitted altogether. What is produced at

Table 2.4 Land-use classification in different countries

	United States	Canada	United Kingdom	The Netherlands
Date of development	1961	1963	1966	1961
Separate systems for:				
Agriculture	√	√	√	√
Forestry	×	√	×	×
Recreation	×	√	×	×
Wildlife	×	√	×	×
Quantification	×	×	√	√
Factors:				
Climate	√	√	√	×
Nutrients	×	×	√	×
Soil	√	√	√	√
Wetness	√	√	√	√
Erosion	√	×	×	×
Pattern	×	×	√	×
Vegetation	×	√	√	×
Topography	√	√	√	√

the end of the process is essentially a classification indicating the suitability of the site to produce crops, with an indication of what particular factors are limiting, which can be applied at field scale. Canada also has a system of soil classification for reclamation purposes.

What follows is a detailed description of two of the systems in use in the United Kingdom.

2.4.1   Agricultural Land Classification

In the United Kingdom a comprehensive Agricultural Land Classification (ALC) exists based on a survey carried out between 1966 and 1974 (Ministry of Agriculture, Fisheries and Food, 1974). This classification focuses on suitability of land for the production of crops and is based on three principal factors: soil, topography and climate. Other factors such as management practices, fixed equipment and farm structure are deemed less important as they may be regarded as transitory. The physical limitations may operate in a number of ways by:

1.   Affecting the range of crops which may be grown.
2.   Influencing yield.
3.   Determining the consistency of yield.
4.   Affecting the cost of gaining the yield.

This classification may be based on the experience of the Agricultural Development and Advisory Service (ADAS) officer and the farmer and with reference to the production figures and costs, but in the final analysis relies, in part, on application of some subjective criteria.

There are five grades of land.

Grade I Land with very minor or no physical limitations to agricultural use. Grade I land is deep, well-drained loams, sandy loams, silt loams or peat, lying on level ground or gentle slopes which are easily cultivated. It has good water-holding capacity or water table within the reach of roots, giving good reserves of water, and is well supplied with nutrients or readily responsive to fertilizer application. There is no major restriction by climatic factors. Yields are consistently high and cropping highly flexible, because most crops may be grown on such land, including horticultural crops requiring rigorous management.

Examples may be found on loamy alluvial deposits, drained lowland peats and reclaimed estuarine and marine deposits.

Grade II Land with some minor limitations which exclude it from Grade I, often connected with the soil; e.g. texture depth or drainage. Minor climatic or topographic factors may also cause it to be included, such as exposure or slope. These limitations may hinder cultivations and harvesting, leading to lower yields and flexibility. A wide range of crops may still be grown, through opportunities for growing arable root crops and some horticultural species.

Examples may be found amongst loess over or intermixed with clay-with-flints, siltstones, chalk and limestone, but all mainly loams.

Grade III Land with moderate limitations owing to the soil, topography or climate, or a combination of these. These limitations lead to a restriction in the choice of crops, timing of cultivations, or yields. Soil defects may include structure, water-holding capacity, texture, drainage, depth or stoniness. The range of cropping is fairly restricted on this grade of land. Only the least demanding horticultural crops may be grown, and forage root crops. The principal crops grown are therefore grasses and cereals. Such crops result in top quality permanent grassland in this grade, where arable cropping is hindered by physical restrictions (e.g. slope and erosion risk).

Examples may be found amongst boulder clays, clay vales (e.g. London Clays), poorly drained estuarine or marine deposits, shallow soils on limestone (<25 cm (10 in)), sandstones and sandy drifts. Grade III may be subdivided into classes a, b and c.

Grade IV Land with severe restrictions, owing to unsuitable texture, poor structure, wetness, shallow depth, stoniness and low water-holding capacity. Relief and climatic restrictions may include steep slopes (between 1 in 5 and 1 in 3), short growing season, high rainfall (land over 200 m and >130 cm (600 ft and >50 in) per annum) and exposure. Land in this grade is generally only useful for low-output uses, such as grass, sometimes with oats, barley and forage crops.

Examples include land fringing uplands, poorly drained areas of vales, hard rocks weathering to shallow and stony soils, and alluvium subject to regular flooding.

Grade V Land of little agricultural value, with severe limitations because of very steep slopes, excessive rainfall and exposure, poor to very poor drainage, low water-holding capacities and severe plant nutrient deficiencies and sometimes toxicities. Land over 300 m (1000 ft) with more than 150 cm (60 in) of rainfall per annum, or with steep slopes (> 1 in 3). Such land is grass or rough grazing; very occasionally forage crops may be grown.

Examples are found in wetland areas impossible to drain, undrained peats, mountain and upland grazing and *polluted areas*.

This system is focused entirely on agriculture and gives no regard to conservation resources, such as woodlands, wetlands and bogs. Indeed in the above classification these areas would have a grading of 5, the same as polluted areas! The farmer has, in the past, been given financial assistance to fertilize, drain and 'clear' areas rich with flora and faunas, in order to 'improve' the agricultural grade. This has led to massive losses in wetlands (Somerset Levels), species-rich grasslands and many woodlands. Clearly this system is of limited use to the restoration ecologist, but has utility in application to agricultural reclamations.

2.4.2   *Land-use capability classes*

Land-use capability classes are the system devised by the Soil Survey of England and Wales, based on the system devised by the US Department of Agriculture (USDA) Soil Conservation Service. The principal distinguishing characteristic from the ALC system is that this is a *capability* assessment, and assumes *a moderately high level of management* and not *present use*. Also the capability assessment may be affected by major reclamation works, such as pumping schemes, which permanently change the previous limitations. The classification also emphasizes physical rather than chemical limitations, as the latter are more easily overcome with modern fertilizers.

There are seven land-use capability classes.

Class 1   Land with very minor or no physical limitations to use Such land is well-drained deep loams, sandy loams or silty loams, related humic variants or peat, with good reserves of moisture or with suitable access for roots to moisture. They are either well supplied with plant nutrients or responsive to fertilizers. Such land is level or gently sloping with a favourable climate. A wide range of crops is possible, with good yields from moderate fertilizer use.

Class 2   Land with minor limitations that reduce the choice of crops and inter-fere with cultivations   Limitations may include moderate or imperfect drainage, less than ideal rooting depth, slightly unfavourable soil structure and texture, moderate slopes, slight erosion and slightly unfavourable climate. These may occur singly or in combination. A wide range of crops may be grown, but some winter crops and root crops may be difficult due to adverse harvesting conditions.

Class 3   Land with moderate limitations that restrict the choice of crops and/or demand careful management   Limitations arise from imperfect to poor drainage, restrictions in rooting depth, unfavourable structure and texture, strongly sloping ground, slight erosion and moderately unfavourable to severe climate. These affect timing of cultivations and range of crops; restricted mainly to grass, cereal and forage crops. Good yields are possible but difficulties are more difficult to overcome.

Class 4   Land with moderately severe limitations that restrict the choice of crops and/or require very careful management practices   Limitations due to poor drainage are difficult to remedy, and there may be occasional damaging floods, shallow and/or stony soils, moderately steep gradients, slight erosion and a moderately severe climate. These combine to limit the range of crops possible and increase the risk of crop failure. Main crop grass, with cereals and forage crops as alternatives, if somewhat risky.

Class 5   Land with severe limitations that restrict its use to pasture, forestry and recreation   Limitations are due to one or more of the following defects which cannot be remedied: poor or very poor drainage, frequent damaging floods, steep slopes, severe risks of erosion and severe climate. Arable cropping is prohibited on this land class, although mechanical pasture improvement is possible. Wide capability for forestry or recreation.

Class 6   Land with very severe limitations that restrict use to rough grazing, forestry and recreation   Such land has one or more limitations that cannot be corrected: very poor drainage, liability to frequent damaging floods, shallow soil, stones or boulders, very steep slopes, severe erosion and very severe climate. The limitations are severe enough to prevent the use of machinery. Very steep ground which is grazable on a long-term basis is included, as are wet areas on level or gently sloping sites.

Class 7   Land with extremely severe limitations that cannot be rectified   Limitations are as a result of one or more of the following: poorly drained boggy soils, extremely stony, rocky or boulder strewn soils, bare rock, scree, or beach sand or gravels, untreated waste tips, very steep gradients, severe erosion and extremely severe climate. Exposed situations, protracted snow cover, and a short growing season preclude forestry, although some rough grazing may be possible.

Sub-classes   These are used to indicate the type of limitation affecting land-use: w (wetness), s (soil limitations), g (gradient and soil pattern limitations), e (liability to erosion) and c (climatic limitations).

Thus a fairly detailed assessment of a site may be made which may be used as a benchmark for the restoration or reclamation process. However, we are again faced with the fact that this classification system has little scope for including *natural* or *semi-natural* ecosystems in *prime locations*. Still less is there scope for inclusion of integrated, functional landscape process considerations.

2.4.3   Protected areas

There is another type of land classification designed specifically for the protection of natural areas. Currently there are a number of types of protection available, with varying degrees of statutory basis and control, including local nature reserves, Sites of Special Scientific Interest (SSSIs), National Nature Reserves (NNRs), and a variety of local wildlife trust areas. On a global scale there are the World Heritage Sites which fall under the aegis of the United Nations Educational, Scientific and Cultural Organization. The factors used to accord such status of these sites varies from nation to nation and, especially, locality to locality. Principally the presence of rare or endangered species will be a high priority, but there are also considerations of landform, geological features and cultural importance. Different nations and local groups may apply different weights to these factors, so at this stage it is probably counter-productive to be prescriptive.

These areas are subject to a wide range of planning controls which must be taken into consideration when planning a disturbing or contaminative use, but do not necessarily guarantee permanent protection. Between 1984 and 1988 over 680 SSSIs (about 14 per cent of the total) had suffered damage.

2.4.4   United States Geological Survey (USGS) Land Classification System

The USGS has developed a two-tier classification system based on present land-use as determined by remote-sensing by satellite and aerial photography. Six of the nine classes are biologically based, and the system is comprehensive in this respect (Table 2.5). Although not all of these categories apply to the United Kingdom, it is possible to see how they could be applied, and extended to a global classification system, allowing ecosystems to be included in the system.

Table 2.5   United States Geological Survey Land Classification System

Level I	Level II
1 Urban or built land	11 Residential 12 Commercial and services 13 Industrial 14 Transport, communication and utilities
2 Agricultural land	21 Cropland and pasture 22 Orchards, groves, vineyards, nurseries, and ornamental horticultural areas 23 Confined feeding operations 24 Other agricultural land
3 Range land	31 Herbaceous rangeland 32 Shrub-brushland, rangeland 33 Mixed rangeland
4 Forest land	41 Deciduous forest land 42 Evergreen forest land 43 Mixed forest land
5 Water	51 Streams and canals 52 Lakes 53 Reservoirs 54 Bays and estuaries
6 Wetland	61 Forested wetland 62 Non-forested wetland
7 Barren land	71 Dry salt flats 72 Beaches 73 Sandy areas other than beaches 74 Bare exposed rock 75 Strip mines, quarries and gravel pits 76 Transitional areas 77 Mixed barren land
8 Tundra	81 Shrub and brush tundra 82 Herbaceous tundra 83 Bare ground tundra 84 Wet tundra 85 Mixed tundra
9 Perennial snow/ice	91 Perennial snowfields 92 Glaciers

2.5   Environmental audit and impact assessment

Environmental impact assessment is a formalized procedure for assessing the impacts (good or bad) of a development or project, on human welfare and/or the environment. It may be used to determine whether statutory requirements will be met, or whether the development will be deemed environmentally acceptable. As so often is the case for environmental action and legislation, the first country to formalize the role of the EIA in the planning procedure was the United States, as a result of the National Environment Policy Act (NEPA) 1969.

In Europe in the early 1980s despite resistance from some EU member states a draft directive was adopted and finally became law in 1988. It is left to member states as to exactly how the legislation will be implemented.

2.5.1   The process of EIA

This consists of several steps:

1.   Proposal.
2.   Consultation.
3.   Project outline.
4.   Scoping.
5.   Interactions.
6.   Modification.
7.   Baseline surveys.
8.   Post-development monitoring.

This process involves a number of groups; the landowner, the developers, the local authority, construction/development engineers, management and assessment professionals and 'interested parties'. This last should include local residents, wildlife and recreation groups and individuals. This process may consist of several iterations at each stage, and indeed the process may be reiterated as a whole several times.

Beanlands and Duinker (1984) suggested that an EIA be required:

1.   To identify early on an initial set of valued ecosystem components to provide a focus for subsequent activities.
2.   To define a context within which the significance of changes in the valued ecosystem components can be determined.
3.   To show clear temporal and spatial contexts for the study and analysis of expected changes in the valued ecosystem components.
4.   To develop an explicit strategy for investigating the interactions between the project and each valued ecosystem component, and demonstrate how the strategy is to be used to co-ordinate the individual studies undertaken.
5.   To state impact predictions explicitly and accompany them with the basis upon which they were made.
6.   To demonstrate and detail a commitment to a well-defined programme for monitoring ecosystem effects.

Morrisey (1993) put more succinctly that the EIA needed to have three stages:

1.   A description of the environment to be disturbed.

2. Acquisition of information as to the nature of the disturbance.
3. Prediction of the likely effect of the disturbance.

Ideally, at the end of the process not only should the impact of the proposed development be fully understood as to its implications for ecosystem function and the socio-economic background, but the groups should all have a stake in the development being both *successful* and *sustainable*. All too often the EIA process is compromised by inadequate consideration of the ecological components, and of the needs and aspirations of the local community.

2.6   Integrated approaches

From what has been discussed so far it is apparent that the approaches to assessment of sites for the change in land-use are thorough when aimed at a specific end. In some respects, however, they are incomplete for a variety of reasons. Most commonly these are questions of scale (off-site effects not being taken into account), incomplete inventory (ecological or other technical parameters being ignored) and failure of analysis (inappropriate techniques being employed). So how might this be overcome? One way that is emerging as being of great utility is the wide-scale adoption of GIS. What must be included is the full-scale integration of scale, time, structure and ecological functions. Some attempts are currently being made to accomplish this and these are outlined below.

2.6.1   Long-term and large-scale approaches

There are a number of programmes proceeding at the moment attempting to address some of the problems of scale in space and time.

Land Condition Trends Analysis   Recently the United States Army has developed a standard method of data collection, analysis and reporting called the Land Condition Trend Analysis (LCTA) Programme (Deirsing *et al.*, 1992). This method utilizes audits of vascular plants, and wildlife inventories, on permanent field plots (Fig. 2.4). This integrates satellite imagery with ground proof data by means of GIS. Ground cover, canopy cover, woody plant density, slope length, slope gradient, soil information and disturbance data are collected.

The Oregon Transect Ecosystem Research Project (OTTER)   This project, described by Peterson and Wareing (1994), is based in the Pacific North West of the United States, and involves six major sites from the interior to the coast. A number of parameters are measured such as the leaf area index, derived from satellite image data. This has enabled determination of the effects

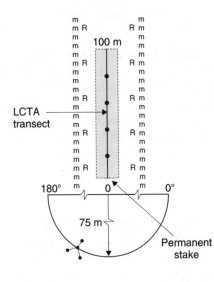

Fig. 2.4 Diagram of the US army LCTA programme permanent field plot. Shaded area is the 600 m² belt transect. R and m are rat and mouse trapping sites. The arc at the end of the plot represents the potential location of the amphibian and reptile trapping sites.

of certain management practices on ecosystem parameters and the development of a predictive model. Although primarily designed to investigate the possibility of developing accurate models from satellite images, it has clearly demonstrated the potential of applying this approach more widely.

These kinds of approach are comprehensive and offer a way forward in assessing the intrinsic 'value' of a site in far broader than economic terms.

2.6.2 Ecological soundness

There are several features of the ecosystem under study which need to be quantified in order to gain an indicator of its ecological soundness; in other words, that which reflects the system's function, structure and stability. These include hydrosphere, lithosphere, biosphere and atmosphere.

Determination of ecological soundness will involve measurements of nutrient cycling, hydrology and resource conservation. In 1988 the US Environmental Protection Agency (USEPA) established the Environmental Monitoring and Assessment Programme (EMAP), which includes a full examination of these factors (Kutz and Linthurst, 1990). The EMAP contains a full integration of the following factors, amongst others:

1. Soil integrity including soil structure, water- and nutrient-holding capacity, vulnerability to erosion, extent of acidification, salinization and contamination, and biological components.
2. Chemical exports from areas.
3. Water availability, quality and run-off.
4. Wildlife habitat quality.
5. Land-use patterns.

Unlike the ALC and Land Use Capability Classification (LUCC) employed in the United Kingdom, the full range of land-use, including natural and semi-natural areas, may be assessed and there is a high degree of *quantitative assessment*. This could be changed if there were integration of the National Vegetational Classification with measurements of soil biological and functional components, and the Soil Survey and Land Research Centres GIS. If the prediction capabilities of the FIBS programme of the UCPE at Sheffield were added, we would have an extraordinarily powerful tool for predicting the impacts of land-use, and for providing prescriptions for effective and efficient restoration.

There is a suggestion that the concept of 'ecosystem health' might be adopted as the final goal of environmental management and assessment (Shrader-Frechette, 1994). The definition of a healthy ecosystem is one which can maintain desirable vital signs in the face of environmental stress and can recover equilibrium after perturbations. Although it has been argued that this definition is too dependent on a value judgement without a clear set of operational parameters, this is hardly a reason for rejecting a potentially useful paradigm for study and action, and deserves further attention and development.

2.6.3   Integration of ecological and economic measures

The final piece in the jigsaw of analysis and interpretation is that of the socio-economic factor. Although this is included implicitly in the EIA process, there are currently no reliable equations for linking economic variables with ecosystem characteristics and dynamics. Therein lies a significant challenge to scientists and economists, not least in communicating with each other in the first instance. There are, however, several examples of efforts to develop such approaches (Barde and Pearce, 1991).

Unsworth and Bishop (1994) have pointed out that the cost of producing a definitive economic value on the loss of wetland 'ecosystem services' exceeds that of the expected damages. They have proposed a simpler model based on the provision of 'environmental annuities': in other words, the provision of the same service with the same effect, but elsewhere, and the authors have demonstrated the feasibility of this approach. Although this has a certain intellectual attraction, it is a means of allowing unwarranted ecosystem destruction to proceed. If ecological restoration is about anything, it is about questions of place, i.e. what is happening here?

Stomph *et al.* (1994) have proposed a system for the integration of bio-physical parameters and socio-economic sub-systems, in the context of cropping practices of villages in the developing world. This certainly looks promising, and may provide a framework for future work.

2.7 Conclusions

It is clear that the technology already exists for the adequate assessment of resources at a number of scales, in space and time. What is now required is their integration at a number of levels, such that when there is to be a development or restoration planned then there will be due consideration of its effects at all levels of parameter type and social acceptability.

References

BARDE, J-P. and PEARCE, D. W. (1991). *Valuing the environment: six case studies*. Earthscan Publications, London.

BEANLANDS, G.E. and DUINKER, P.N. (1984). An ecological framework for environmental impact assessment. *J. Environ. Man.*, **18**, 267–277.

BENTHAM, H., HARRIS, J.A., BIRCH, P. and SHORT, K.C. (1992). Habitat classification and soil restoration assessment using analysis of soil microbiological and physico-chemical characteristics. *J. Appl. Ecol.*, **29**, 711–718.

DEIRSING, V.E., SHAW, R.B. and TASZIK, D.J. (1992). US Army Land Condition Trend Analysis (LCTA) Programme. *Environ. Man.*, **16**, 405–414.

GOLDSTEIN, R. and FERSON, S. (1994). Response of plants to interacting stresses (ROPIS): program rationale, design, and implications. *J. Environ. Qual.*, **23**(3), 407–411.

GRIME, J.P., HODGESON, J. and HUNT, R. (1988). *Comparative plant ecology: a functional approach to common British species*. Unwin Hyman, London.

HUND, K. and TRAUNSPURGER, W. (1994). Ecotox – evaluation strategy for soil bioremediation exemplified for a PAH-contaminated site. *Chemosphere*, **29**(2): 371–390.

KLIJN, F. and DE HAES, A.U. (1994). A hierarchical approach to ecosystems and its implications for ecological land classification. *Landscape Ecol.*, **9**(2), 89–104.

KRABBENHOFT, K., KIRBY, D., BIONDINI, M., HALVORSON, G. and NILSON, D. (1993). Topoedaphic Unit Analysis: A site classification system for reclaimed mined lands. *CATENA*, **20**, 289–301.

KUTZ, F.W. and LINTHURST, R.A. (1990). A system-level approach to environmental assessment. *Toxicol. Environ. Chem.*, **28**, 105–114.

MINISTRY OF AGRICULTURE, FISHERIES AND FOOD: AGRICULTURAL DEVELOPMENT AND ADVISORY SERVICE (1974). *Agricultural land classification of England and Wales*. HMSO, London.

MORRISEY, D.J. (1993). Environment impact assessment – a review of its aims and recent developments. *Marine Poll. Bull.*, **26**(10), 540–545.

O'RIORDAN, T. (1971). *Perspectives on resource management*. Pion, London.

PERRING, R. H. and FARRELL, I. (1983). *British Red Data Books 1, Vascular plants*, 2nd Edn. RSNC, Nettleham.

PETERSON, D.L. and WAREING, R.H. (1994). Overview of the Oregon Transect Ecosystem Research Project. *Ecol. Applications*, **4**(2), 211–225.

PETERSON, R.L. and FARQUHAR, M.L. (1994). Mycorrhizas – integrated development between roots and fungi. *Mycologia*, **86**(3), 311–326.

READ, D.J. (1992). The mycorrhizal mycelium. In Allen, M.F. (ed.) *Mycorrhizal functioning: an integrative plant–fungal process*, pp. 102–133. Chapman and Hall, New York and London.

RODWELL, J. (ed.) (1989 *et seq.*). *British plant communities*. Cambridge University Press, Cambridge.

SHRADER-FRECHETTE, K.S. (1994). Ecosystem health: a new paradigm for ecological assessment. *Trends Ecol. Evol.*, **9**(12), 456–457.

SPELLERBERG, I.E. (1991). *Monitoring ecological change*, pp. 79–80. Cambridge University Press, Cambridge.

STOMPH, T.J., FRESCO, L.O. and VAN KEULEN, H. (1994). Land use system evaluation: concepts and methodology. *Agric. Syst.*, **44**, 243–255.

UNSWORTH, R.E. and BISHOP, R.C. (1994). Assessing natural resource damages using environmental annuities. *Ecol. Econ.*, **11**, 35–41.

WRIGHT, J.F., FURSE, M.T. and ARMITAGE, P.D. (1993). RIVPACS – a technique for evaluating the biological quality of rivers in the UK. *Europ. Water Poll. Control*, **3**(4), 15–25.

Further reading

DAVIDSON, D.A. (1992). *The evaluation of land resources*. Longman, Harlow.

Reclamation of land

Chapter 3

Green field sites

3.1 Introduction

In the not so distant past it would have been unusual to include a section in a book of this sort on sites prior to disturbance. The process of restoration or reclamation is normally only considered after the industrial use or agro-forestry had been completed. It is a sign of the maturity of this subject that it now recognizes that restoration or reclamation planning should begin *prior to change of land-use*, as outlined in Chapter 1. In most cases, this will still be agriculture → disturbance → agriculture, but other pathways are now emerging, such as agriculture → disturbance → conservation. In either case, it is essential that the state of the site be carefully assessed prior to any work beginning. Indeed, under current legislation, it is impossible to begin work legally without a full consultation and planning procedure. In this chapter we will consider the changes occurring on sites affected by civil engineering operations, agro-forestry and recreation. Areas contaminated with the by-products of industry are a special case and will be considered in Chapter 4.

This chapter draws largely on experience gained from investigating areas subject to open-cast mining, but the changes illustrated caused by soil movement, storage and reinstatement will apply to other uses such as roads and pipelines.

3.2 Activities prior to soil movement

There are four stages in the change in land-use necessitating a reclamation or restoration programme: prospecting, planning, use and rehabilitation. Rehabilitation, and all other post-use activities, will be considered in later chapters.

3.2.1 Prospecting

Prospecting can range from a company wishing to identify a green field site upon which to build a new superstore to a mining company investigating the extent and quality of mineral deposits. The former type of prospecting is carried out by means of interrogation of GIS databases, which can be asked sophisticated questions such as 'which area in County X has a population density of Y not yet served by a superstore or DIY outlet?' This can very quickly identify a good location. In the case of mining, the procedures take longer as, although a general geological survey map may exist for the area, the reserves will still need to be 'proven' by drilling for cores and determination of the hydrological nature of the site. Although the United Kingdom has a largely well-characterized geology, this is not the case in all countries and, in many areas, large tracts of land remain undisturbed as the reserves of minerals available are unknown. Prospecting for mineral reserves was one of the principal activities of European empires in the Third World. This activity may be better termed *exploration*.

3.2.2 Planning

In order to use a site in the United Kingdom the company wishing to remove the mineral or have temporary access for other purposes may purchase the land completely, or draw up an agreement with the landowner, allowing the works to be carried out. This agreement will consist of an annual rental to the land-owner, and some compensation for the loss of profit that the landowner would have accrued during the use of the land. The payment would be for all of the land taken out of production or other use, for example area not actually to be used for the activity directly, such as parking and office areas. The costs of vacating the land, disposal of livestock, re-stocking after the use has finished and any decrease in market value are also included.

At the same time, before work can proceed, a detailed environmental impact assessment needs to be prepared (cf. Chapter 2). Then a period of consultation is entered into with the local authority, statutory bodies and other interested parties (Table 3.1). This statement is modified following consultation, and a restoration or reclamation plan devised. The plan will include: full details of what will happen to soil resources, how waterways will be protected, details of contours to be reinstated, drainage plans, conservation of on-site biological resources, farm-fixed equipment and biological management. The planning application is then made formally and may be accepted. If so, then the work may proceed, the plans may be abandoned (very unusual) or more likely, an appeal lodged resulting in a public inquiry. Similarly, if the company is successful in its application but a local body or individual dissents, leave to appeal against the permission may be granted; again resulting in a public inquiry.

Table 3.1   Interested parties to be consulted in England during proposed changes in land-use

Status of consultee	Identity
Local authority	Chief Executive, County Council
	Chief Executive, Borough/District Council
	Chair, Parish Council
	Planning Department
	Engineer and Surveyor
	County Archaeologist
Statutory body	Ministry of Agriculture, Fisheries and Food
	Agricultural Development and Advisory Service
	National Rivers Authority
	Local water company
	British Telecom
	British Rail/Railtrack
	Local electricity company
	National power company
	Forestry Enterprise
	Countryside Commission
Other organization	National Farmers' Union
	Landowners
	Lessees/tenants/occupiers
	Council for the Protection of Rural England
	Church Commissioners
	English Nature
	English Heritage
	National Trust
	Ordnance Survey
	Post Office
	Pipelines Authority
	Footpath societies
	Wildlife trusts
	Health and Safety Executive
	Mining Records Office
	Department of the Environment
	Department of Energy
	Department of Transport
	Department of Trade and Industry

3.3   Soil movement and care

3.3.1   Soil characteristics

One of the principal considerations of any land reclamation or restoration is that of soil quality. Consequently, it is important that if the land-use activity leading to the need for restoration involves soil disturbance, the soil should be handled with great care.

Historically, ALC land in Grades 1, 2 and 3a have been excluded from permission to working by open-cast extraction methods. Now, however, there is great pressure on this type of land as reserves under the other grades have been worked out with respect to certain minerals (notably sand and gravel in South East England). This has also meant that the soils which have been worked are unlikely to be returned to a higher grade of classification after the reclamation process has been completed.

Traditionally, the characteristics taken into account when handling soil have all been based on physical parameters: depth of soil, distribution, horizonation, texture, structure, bulk density and compaction, drainage and plastic limit (Ramsay, 1986). Each of these characteristics defines an important component of the substrate which needs careful treatment and for most purposes may be referred to as *topsoils*, *subsoils* and *overburden* (material below rooting depth but above the economic mineral). We will go on to consider which other characteristics *should* be taken into consideration but which currently are not.

Depth and distribution area   A simple calculation from these two factors gives the total volume of soils available on-site. Because it is impractical to ascertain the actual depth of soil at every point of the site, certain assumptions have to be made as to depth between sampling points, to give a mean value.

Horizonation   This indicates the division of the total volume of soil into the categories of topsoils, subsoils and overburden; these will require separate handling. Subsoils are sometimes divided further into lower and upper horizons, dependent upon the distribution of characteristics within them, such as textural classes and stoniness.

Textural class   This is important to ascertain as it is responsible for determining many physicochemical parameters such as drainage, water-holding capacity and cation-exchange capacity (cf. Chapter 2). Textural class will also have significant consequences for the type and intensity of drainage schemes to be employed, as finer textured soils (i.e. clayey) will require greater attention than sandy soils.

Structural status   This refers to the arrangement of particles within the soil, as described by aggregation, ped structure and stability. Structural status is almost entirely dependent upon the action of the biological components of the soil and the organic matter which arises from it. It is unfortunate, therefore, that so little attention is paid to the biological component in civil engineering procedures.

Bulk density   This describes the amount of solid material, excluding liquid, present per unit total volume. Bulk density is of great importance in determining the potential for a soil to support plant growth. It may vary naturally from between 1 and $1.8\,\mathrm{g\,cm^{-3}}$ depending on texture and stability. In very compact soils, plant roots will be unable to penetrate, as hydrostatic forces tend to work radially from the root axis, rather than along its length. It is in bulk density that the most obvious physical changes to soils during handling may be found, accompanied by loss of structure. The bulk density of the various horizons needs to be recorded in order to provide targets for the reclamation process.

Drainage In the context of a site undisturbed by soil movement (as opposed to the cultivations found in agriculture) the rate and pathways of water drainage is a complex phenomenon resulting from the interaction of several factors: structure, porosity, relief, geohydrology and climate. Generally, the better drained a soil is, the easier it will be to handle.

Plastic limit When considered as an engineering material soil has a point at which it changes from a non-plastic material and begins to act as a plastic, in common with many other cohesive materials. Plastic limit is linked to a measurable moisture content, and is specific for each soil type. Generally, the finer the texture of the soil, the lower its plastic limit will be. If exceeded when being handled, the soil will be extremely susceptible to compaction and smearing. It is for this reason that soils should only be moved when dry, and why the plastic limit is the principal factor taken into consideration when decisions on soil handling are made.

3.3.2 Methodology of soil movement

Lifting The type of equipment used in moving soil (Table 3.2) will be determined by a combination of factors:

1. The area of site and amount of material to be moved.
2. The nature of the soil.
3. Equipment availability in the area or to the contractor.
4. The timescale of the operation.

On large sites the commonest type of equipment used to move soil is the *box earthscraper*. This is a very heavy and powerful vehicle, which has been the workhorse of soil-moving operations on open-cast coal-mines and other civil

Table 3.2 Equipment available for moving soil (Ramsay, 1986)

Lifting	Transport	Placement
Bulldozer	Bulldozer	Bulldozer
Front loader	Dump truck	Light bulldozer/backhoe
Face shovel	Dump truck	Light bulldozer/backhoe
Dragline	Dump truck	Light bulldozer/backhoe
Backhoe	Dump truck	Light bulldozer/backhoe
Front shovel	Dump truck	Light bulldozer/backhoe
Bucket wheel	Conveyor	Light bulldozer/backhoe
Earthscraper	Earthscraper	Earthscraper/backhoe

engineering operations since the Second World War. There is a large box in the centre of the vehicle, which has a blade at its leading edge. This is lowered into the soil in order to cut and lift it into the box. Strips 2–3 metres wide and 5–8 metres in length may be lifted at one time. This equipment exerts a static force of some $8\,kg\,cm^{-2}$ in the wheelings, and there may be momentary peaks of significantly greater amplitude as a result of bouncing. This, combined with the action of the scraper blade, leads to great compaction and shearing forces being exerted on the soils being moved.

The *backhoe* or back acting digger is used for moving smaller volumes of soil, or where a larger budget and more time are available. The bucket can move between 1 and $5\,m^3$ at a time and is less disruptive to soil structure. This may be used directly for transport over a very limited distance, no more than a few metres, or more commonly it is transferred to another vehicle, such as a dump truck.

Transport   The box earthscraper is able to transport the removed soil directly to its new location, be that for direct placement or storage in bunds, although it does have to cross the bund to dump and load. That soil excavated by digger needs to be transported by dumper trucks. At the receiving site the soil may then be removed by tipping or by another backhoe.

Storage   Stores are most commonly constructed by earthscraper, as outlined above. However, there are operations where less destructive equipment may be used. Principally, backhoes may be used to lift soils gently and place them into dumper trucks where they are tipped into shallow dumps. In this way trafficking is kept to a minimum, as is damage. However, even this gentle handling may result in profound changes in soil characteristics.

The size and disposition of these stores is determined by a number of factors. Principally the amount of soil to be stored and the space available on which to build it. Commonly, stores are 3–5 metres high. Stores may be no more than 10 metres in length, but many stretch for 500–600 metres, often around the perimeter of the site, acting as an environmental baffle and affording some degree of protection for the outside of the site from noise and dust nuisance.

3.3.3   *Changes in soil as a result of handling*

The principal changes to soil are decreases in soil porosity due to the compaction and shearing forces imparted during handling. In directly replaced soils there may be a decrease of soil bulk densities of up to 40 per cent, representing almost all of the free air spaces. This has a severe effect on the ability of plant roots to penetrate, and compromises water and gas movement. Also, there is much damage to living organisms, with large decreases in earthworm populations. This can also result in the release from dormancy of buried seeds of weed species, from the combination of release of nutrients as a result of death of biomass, and the exposure to light.

3.4 Microbiological, chemical and physical changes in soil during storage

It is during the storage of soil that the most profound changes occur, to the detriment of its function as a substrate for plant growth and development.

3.4.1 Physical changes

Although the majority of damage to soil structure is from handling methods, it is further exacerbated during the storage period (Abdul-Kareem and McRae, 1984; Scullion, 1994). This is due to the breakdown of organic materials and biological structures binding soil particles together. As the soil is made increasingly compact, the volume occupied by gas and available water declines rapidly (Fig. 3.1). A loss in aggregate stability results, as was demonstrated by Scullion (1994; see Fig. 3.2). Loss in aggregate stability represents a large decrease in the ability of the soil to provide adequate aeration for plant growth and drainage. Any channels which are formed either by the action of soil invertebrates or by artificial means will be quickly blocked by fine soil particles no longer bound into stable aggregates.

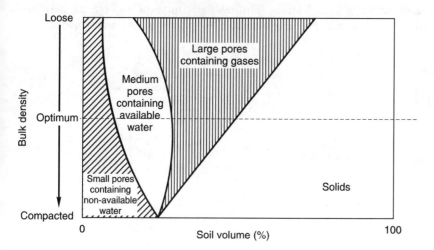

Fig. 3.1 Composition of soils at different degrees of compaction (from Wilson, 1985).

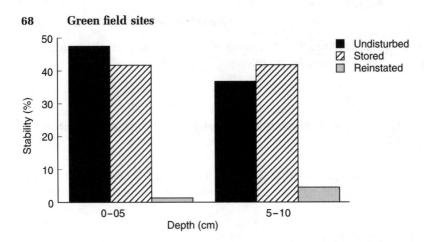

Fig. 3.2   Stability of silt and clay fractions for soils at various stages of mining operations.

3.4.2   Chemical changes

Initially there is a decrease in the availability over the first few months of storage, as the organic materials are oxidized biologically. The supply of oxygen to the soils has been cut off due to compaction, leading to the development of anaerobic conditions during the first three months. At the same time oxidized inorganic compounds, such as nitrates, are also reduced until only reduced compounds are found. This leads to a number of important changes in the soils' chemistry. The lack of oxygen prevents the transformation of organic nitrogen to nitrate, during mineralization. As a result, the process stops at the production of ammonium, which begins to accumulate (Fig. 3.3), and nitrate/nitrite declines. During the course of storage most of the nitrogen which would become available to plants in the undisturbed soils

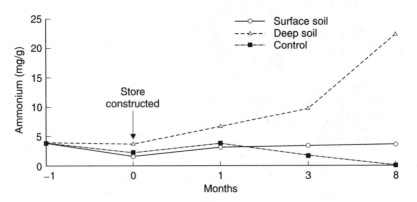

Fig. 3.3   Accumulation of ammonium during soil storage.

accumulates in this form. Nitrogen accumulated is then liable to be lost on re-spreading. Organic acids such as succinate and lactate begin to accumulate in the store.

The change in redox potential is accompanied by a fall in pH (partly engendered by the production of organic acids) and an increase in the *availability* of certain metals, such as manganese, copper and zinc. These can bring further problems of toxicity for the surviving organisms.

Sulphides are then produced as a result of anaerobic metabolism, resulting in the production of metal sulphide precipitates and giving the soil a characteristic dark grey or black coloration.

3.4.3 Biological changes

There are significant and rapid changes in the living component of the soil during the course of storage. It occurs in two major phases as a result of linked phenomena. Initially, because of the large amounts of organic materials, such as fungal hyphae, plant roots and soil animals, killed by the soil lifting and store construction processes, there is a rapid and large increase in the numbers of bacteria throughout the store (Fig. 3.4). This is followed by a period where the numbers of bacteria decline as these reserves are exhausted, until numbers found in reference areas are reached. Numbers of fungi decline immediately, as does total microbial biomass. The fungal biomass is providing organic and inorganic substrates for the bacterial explosion.

The major invertebrates are extremely susceptible to the physical operation of soil movement. They are liable to be crushed and unable to find physical

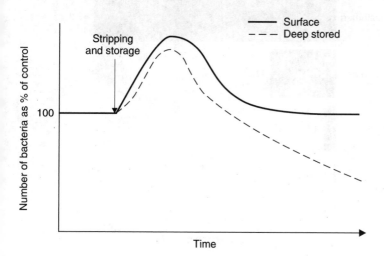

Fig. 3.4 Changes in bacterial numbers as a result of soil movement and store construction.

refuges from the large compaction and shearing forces being imparted to the soil by the heavy earth-moving equipment. This results in the large-scale killing of invertebrates, particularly earthworms, on such sites (Fig. 3.5).

3.4.4 Modelling soil storage

It has become apparent that the conditions within topsoil stores allow division into three primary zones (Harris *et al.*, 1989; Fig. 3.6). The layers at and near the surface remain aerobic throughout the duration of storage, and show some degree of recovery in biological potential and physicochemical conditions after two to three years of storage. At the core of the store, once conditions become anaerobic this continues during the course of storage, bringing the changes outlined above. There is, in between these two, a transitional zone which will fluctuate between the aerobic and anaerobic states, depending on prevailing moisture conditions, i.e. rain will cause the pores to fill and this zone will become anaerobic, but will return to aerobic conditions upon drying out.

The size and speed of the changes in the soil in the store, and the sizes of the three zones described above, are largely controlled by the textural character-istics of the soil. These changes are best illustrated by examining the extent of the anaerobic zone in the soil store (Fig. 3.7). Essentially, the more finely textured the soil, the greater the extent of the anaerobic zone. Consequently, clay soils will become largely (80–90 per cent) anaerobic at the heights to which they are commonly constructed on most civil engineering sites, i.e. 3–4 m. Sandy soils, on the other hand, can be built to these heights with only minimal

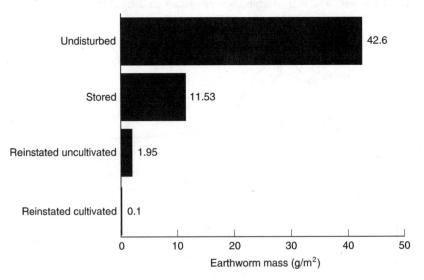

Fig. 3.5 Decline in mass of earthworms as a result of soil storage and cultivation (from Scullion, 1994).

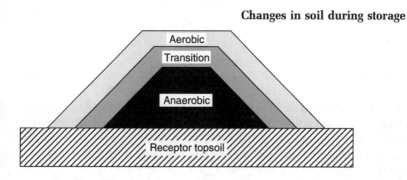

Fig. 3.6   Zonation of soil store.

anaerobiosis setting in. It therefore follows that stores should be kept as low as possible, within the constraints of available space. This must be set against the fact that in so doing, more undisturbed space will be taken up to act as a base upon which to construct the store.

The amount of anaerobiosis which has been present in stored soils should be taken into account when considering what management treatments to apply post-reinstatement, as the soil subject to anaerobic conditions will require greater management inputs than is necessary for direct placement of aerobic zone soil.

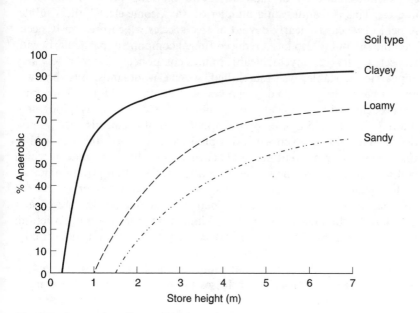

Fig. 3.7   Interactive effects of height and textural class on the size of the anaerobic zone in soil stores (from Harris and Birch, 1990).

3.4.5   Direct placement

One of the ways to avoid many of the problems associated with the long-term storage of soils is to avoid it altogether. This is largely controlled by the size of site available and the speed of working. The soil is stripped and moved directly to a receiving area, obviating the need for storage. On open-cast coal sites this is rarely carried out as there is insufficient working space to receive the new material. On many sand and gravel sites, however, this practice is very common, often with sites in active agricultural use.

3.5   Changes in soil as a result of agriculture or forestry

The use of the land for agriculture and forestry is by far the largest component of those ecosystems affected by human activity. The potential for restoring habitats and species diversity in this cropped landscape remains high. It has been suggested by Bunce and Jenkins (1989), that this could begin at field margins and along linear features. The principal effects of the growing and harvesting of crops are:

1. **Reduction in biodiversity**. It is inherent to the nature of cropping systems that the primary tool of management is the exclusion of all species that can compete, directly or indirectly, with the crop plant or animal of interest. This is manifest in a number of ways, some clear, others subtle. A field of wheat is clearly devoid of tree species which previously occupied the site but it also has a reduced fungal component, particularly with respect to its ectomycorrhizal-forming component. Deer-parks may appear to be stable, well-established, diverse woodlands, but they will almost certainly be devoid of wolves, at least in the United Kingdom. This is achieved by physically removing unwanted organisms (ploughing, hunting, hand-weeding), or by the application of chemicals such as pesticides and poisons. Even when this pressure is removed, this can lead to the presence of persistent and toxic chemicals in the environment, which will have to be overcome in the restoration/reclamation programme.

2. **Soil structure**. As has been described previously (cf. Chapter 1), the soil has a structure which is open and dynamic, maintained by the presence of a metastable biological community. Limited diversity can put structural stability at risk by the combined effects of fertilization and tillage, constantly removing biological potential and reducing the organic material necessary to supply it with nutrition. In most cases this has led to increased rates of erosion, with losses far outstripping the gains made from weathering of new bedrock minerals. Increasingly, there are catastrophic losses as a result of soil mismanagement, primarily in regions where loss of tree cover leads to rapid water erosion and in arid regions,

where inadequate crop management combined with low rainfall leads to wind erosion.

3. **Soil water balance**. It is important that soils are maintained such that crop roots are kept supplied with both water and air. In practice this is achieved by drainage and irrigation. This has the effect of reducing local pockets of variability, more or less waterlogged, where different species may survive.

4. **Fertility**. The manipulation of the fertility of the soil is central to crop management. The majority of crop plants are ruderal in nature, so this is usually achieved by the addition of readily plant-available inorganic nutrients. Use of fertilizers effectively excludes slower growing competitor and stress-tolerating plants, when combined with the cultivation described above. When stopped, the balance of molecules present will still be in favour of the smaller ones such as inorganic ions (cf. Chapter 1, entropy and chaos), which will have to be re-organized into more complex structures, such as the desirable plants and animals, for the restoration to be successful.

These changes will not only affect the site directly but will also have effects off-site, by the export of particulate matter, organic material, inorganic molecules and increased pressure from displaced animals.

We will now examine some cases where various facets of these problems have become manifest. Pywell *et al.* (1994) investigated the interaction between the fertility of soils on farmlands and its implications for their restoration to heathland. The farmland was found to have raised pH (from 3.8 to about 5.5) and fertility as a result of farming practices and the cultivation by ploughing had mixed the normally quite separate soil horizons. When allowed to recover, by ceasing farming operations, the concentrations of phosphorus and nitrogen fell, but the pH remained high up to 13 years after abandonment. This reflects the persistent nature of liming as a treatment, and points to the long-term planning required for restoration of such systems. It is likely that there will either have to be a very active programme of acidification, which is not easily brought about, or a planned succession to successively more acid-tolerant species. Such restorations are thought to be of great importance, as the heathland is important for providing food plants for a range of insect species, particularly when fragments can be expanded and rejoined (Webb and Thomas, 1994).

The impact of the fertility of a soil on the species which it will support is a powerful one. Marrs and Gough (1989) reviewed some of the experiments carried out on the Park Grass plots at Rothamsted. There was a general trend to species impoverishment on fertilized plots, except in the case of the addition of a restricted range, with the exclusion of inorganic nitrogen (Fig. 3.8). This impoverishment was particularly marked on the increasing use of nitrogen, where the slower growing stress-tolerant and competitive species were excluded by ruderal species. The same authors also reported distinct differences in the

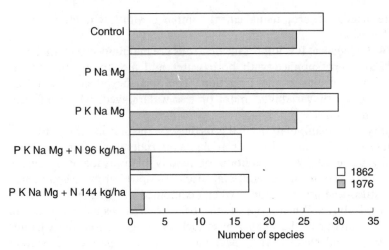

Fig. 3.8 Effects of fertilization on species number on the Park Grass plots at Rothamsted (source, Marrs and Gough, 1989).

amount of phosphorus under different land-uses (Fig. 3.9). Phosphorus excesses were particularly marked in agricultural sites, which had high concentrations of phosphorus, reflecting their fertilizer history. What was also interesting was that sandy soils had particularly high levels, perhaps indicating the lack of uptake in these soils or the over-use of fertilizers to maintain concentrations in crop rooting zones. They concluded that the fertility of soils could be a barrier to the establishment of desirable species on sites undergoing restoration. This phenomenon is not restricted to Northern Europe, as Sparling *et al.* (1994) recorded the highest concentrations of soil phosphorus $(0.88\,\mathrm{mg\,g^{-1}})$, twice that recorded for undisturbed forest areas in New Zealand. There are also some instances where some agricultural practices have increased the biodiversity of some areas. The widespread use of basic slag as a fertilizer in Wales led to species-rich pasture in otherwise acid grassland areas. The decline of the steel industry and the liming option has led to the re-acidification of this land.

Many of these changes occur gradually over a long period of time. Sometimes change in land-use can bring about rapid and dramatic differences in physicochemical conditions. Likens and Borman (1975) monitored the quality of stream water in two forested catchments. One was experimentally deforested in the winter of 1965 and this led, after a lag of around six months, to significant increases in the concentrations of calcium, potassium and nitrate ions, and an increase in sulphate. The former has been attributed to the uncoupling of nutrient cycling, leading to the wholesale export of nutrients from the system. Although no satisfactory explanation was given for the increase in sulphate at the time, recent work in northern Bohemia suggests that this could be due to the cessation of interception of sulphur compounds by the trees, so the compounds no longer settle in the catchment. As to the ecological impor-

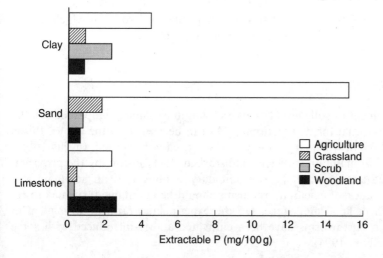

Fig. 3.9 Effects of land-use and soil type on soil phosphorus concentration.

tance of the mechanism of mass loss, which could occur as a result of a hurricane incident, it is difficult to be certain but the lack of readily available nutrients in higher concentrations may suppress the influx of ruderal species and favour the re-establishment of dominant tree species.

3.6 The impact of recreational and amenity uses

The major impacts on recreation are similar to that of agro-forestry, but tend to be more tightly focused in areas set aside for public amenity. Problems arise from misuse of resources, in particular:

1. Erosion of slopes and pathways from intense pressure from walkers and riders.
2. Decreases in diversity by poaching of endangered or key species, or by litter leading to deaths, such as wildfowl being trapped in nylon line.
3. Point pollution incidents, through fly-tipping or careless rubbish disposal.

In many cases there is a need to educate the users of the recreational resource, coupled with the provision of observation facilities and litter bins. Permits for hunting and fishing uses are an essential part of this strategy.

3.7 Conservation of on-site ecological resources

Of crucial importance in all sites where a self-sustaining system is the intended end-use, it is essential that the biological potential of the site be preserved. In some cases there may be organisms unique in the area, some of which will be essential to the re-establishment of ecological function.

3.7.1 Soil animals

The management of soil should be carried out in a manner sympathetic to the organisms essential for its functioning. As can be seen from the above, this is extremely difficult when large volumes of soil need to be shifted. In this case, it is important to maintain sufficient undisturbed stock on-site, by the provision of unworked areas within the site boundary. Failing this, an adequate and appropriate stock of essential organisms should be maintained, such as to re-introduce them at a later stage. On the experimental farm at Bryngwyn in South Wales, earthworms have been cultivated and re-introduced with some success (Scullion, 1994).

3.7.2 Micro-organisms

There is a limit to how much can be done to preserve intact the full composition of the microbial community, as so much unavoidable damage is done to the fungal component as a result of movement and store construction. Cairns (1993) has indicated that it is possible for certain micro-organisms to become locally extinct as a result of the lack of provision of temporarily habitable niches. However, many organisms will survive as spores and more will arrive from off-site when the appropriate plant species have established. It is important, however, to maintain some aerobic stocks of soil, and living plants when considering symbiotic relationships, such as the mycorrhizae, and the aerobic surface of large soil stores fulfils this function.

3.7.3 Plant species

Plant species unique to sites will need to be preserved on-site but if they are extremely rare it is unlikely that permission to work the land will be granted in the first instance. If working is unavoidable, then plants must be stored as seeds, underground storage organs or vegetative stock. There are a number of portable and vehicle-towed devices available for stripping seeds for storage (Morgan and Collicutt, 1994), although such mixed inocula cannot be stored in ideal conditions for all the species it contains. If stored products have been collected in proportions likely to bring about their pre-disturbance numbers, it is unlikely that they will be able to be preserved. In many cases common species are available for re-stocking from commercial growers. Certainly, in the case of reclamation to agriculture or to forestry, well-established sources of supply are available. In the case of species-rich systems there are a growing number of commercial enterprises fulfilling this need. This is certainly the case in the United States and for meadow mixes in Northern Europe.

In order to protect particular trees and shrubs, Carey (1994) has recommended that the following measures be taken:

1. The erection of fixed fencing around the crown spread of trees and shrubs.
2. Prevention of the raising of soil levels around the base of plants above 200 mm from the original, so that feeder roots may remain oxygenated. Failing this, drainage should be ensured to prevent development of anaerobiosis.
3. Prevention of loss of soil from the basal area, and therefore feeder roots.
4. No digging works to be carried out closer than 4.5 m to the base of the trees.
5. In wooded areas all digging to be carried out by hand.

There have also been several suggestions for the maintenance of tree ecosystems subject to cropping. One of the approaches showing much promise is that of contour strip logging. Essentially, trees are harvested on the contour of a slope, parallel to the stream or river below. This is done slowly and progressively up the slope. This allows for the rapid regeneration of the forest, as nutrients from the deforested area (cf. Likens and Borman above) are absorbed by the new vegetation. It is important that there is no strip cutting immediately adjacent to the stream. This latter concept of a buffer zone is also being adopted in Western agriculture in an effort to protect streams from run-off arising from agriculture or forestry. Stocking (1995) has summarized the ways in which both soil and water may be conserved (Table 3.3). It is clear that these practices must be extended before catastrophic collapse occurs in many agroforestry systems.

3.7.4   Animals

Badger setts and other burrowing animals' lairs to remain undisturbed and, most importantly, easy egress/ingress from the site made available *along an existing trail*. Replacement roosting sites for birds and bats, and adequate provision made for food sources.

3.8   Habitat translocation

In recent years attempts have been made to *transplant* or *translocate* habitats. This has been defined as 'the removal and subsequent replacement, usually in a new location, of a complete assemblage of plants and animals, with the aim of maintaining the habitat unaltered in its new location' (Bryne, 1990).

Table 3.3 Techniques for soil and water conservation (Stocking, 1995)

Type of technique	Technique	Description
Tillage practices	Strip tillage	Conditioning soil along narrow strips in or adjacent to seed rows, leaving the rest of the soil undisturbed
	Basin listing, or tied ridging	Formation of contour bunds (earth banks) with constructed banks between bunds. Listing achieved by machinery and tied ridging by hand tools
	Conservation tillage	Technique of light harrowing and retention of crop residues at surface
	Minimum or zero tillage	Use of herbicides, then direct drilling into residues. Very little disturbance to the soil
Land formation techniques	Contour bunds	Earth banks up to 2 m wide across slope to form a barrier to run-off and breaks the slope into shorter segments
	Terraces	Earth embankments and major re-formation of the surface. Three main types: diversion, retention and bench terrace
	Terracettes	Small constructions usually to arrest water
Stabilization structures	Gabions	Stone and rock-filled bolsters to protect vulnerable surfaces, e.g. bridges, culverts
	Gully control dams	Usually constructed of brushwood across a gully

The process involves several stages, setting aside the necessary planning processes described in Chapter 2. Firstly the donor site has to be evaluated as to the composition of the vegetational assemblages, the depth and horizonation of soils, drainage patterns, hydrology, topography and the physicochemical characteristics of the soils. A suitable receptor site then needs to be found, which ideally should have characteristics identical to those of the donor site. In practice this is unlikely to be the case for candidate sites, which can be problematical if there are certain characteristics of the donor site responsible for the presence of species of particular interest, through being endangered or locally rare, or functionally significant. Unfortunately, to make such an assessment is extremely costly and may not be adequately carried out. What is more likely to govern the selection of a receptor site is its proximity to the donor site.

The receptor site then needs to be prepared for receiving donor material. This may simply involve stripping off topsoil, in order to remove vegetation and its propagules, which could compete with the donor material. This also has the effect of removing some of the surface chemical characteristics. In the ideal situation the topography of the site should be altered, along with its drainage patterns, to mimic conditions on the donor site. This is a costly operation and may not always be carried out. As in open-cast mining operations trafficking over the subsoil should be kept to a minimum.

The donor site is then cut into turves which are moved and placed on the receptor site. The turves taken should be as large and thick as is possible, in an attempt to retain the functional and structural integrity of the soil/plant unit, and there should be no gaps between turves. This should only be carried out when the invertebrate population is dormant, in order to minimize damage. Unfortunately, the soil is rarely removed to a depth greater than 45–50 cm, cutting the roots of deeper rooting plants, drainage channels and soil structural units. The disturbance also causes earthworms to burrow deeper into the remaining subsoil, and they are consequently not carried into their new location.

In some cases 'blading' is carried out, which involves the removal of topsoil with its vegetation by earthscraper; in other words, direct replacement as in open-cast coal-mining, but without the disturbance of subsoil and overburden materials. In most cases this is less successful than the use of turves for vegetation establishment.

The majority of sites treated in this manner in the United Kingdom have been grasslands and, to a lesser extent, heathland (Helliwell, 1989). Transplantation of large woodlands is currently an unrealistic proposition, although attempts to move woodland ground floras, and small groups of trees may be feasible (Down and Morton, 1989). It is important that correct conditions of the timing of leaf emergence and canopy closure be met if ground flora transplantations are to be successful.

One of the major drawbacks with this technique is the lack of connection in terms of hydrogeology between the transplanted material and the underlying subsoil. Unless drainage channels, root connections and earthworm burrows linking the two are quickly developed, there will be significant problems with waterlogging and droughtiness. Nevertheless, this technique offers the possibility of reclaiming certain habitat types, where the receptor area was of the same type prior to its change of land-use, and if the project has sufficient funds.

3.9   Protection of water courses

The large disturbances occurring on sites during use will have a major effect on the hydrology of the site and the quality of the water on and from the site. For this reason it is essential that water courses be protected during the working life of sites.

Most sites will have surface water features which need to be preserved as to their hydrologic function. More importantly, with respect to the workings on site, permanent streams and seasonally active channels need to be redirected such as not to result in the void filling with water. The redirection may be achieved by piping when on a small scale, but larger schemes will require the installation of large relief ditches. This is important not only for preserving the hydrologic function of the landscape but also for preserving the connection of upstream and downstream biological communities.

Civil engineering sites, as has been demonstrated above, lead to the large-scale disruption of soil structural stability. This leads to the inevitable consequence that there will be a large volume of particulate matter susceptible to being suspended in surface waters as sediment. This will carry an associated load of organic material and reduced compounds. Consequently, if this is released into local water courses, it will result in the oxidation of this material by micro-organisms; in other words it will have large biological and chemical oxygen demands. This leads to oxygen being stripped from water courses and the death of sensitive invertebrates and fish. Therefore the local water regulator (the National Rivers Authority) lays down strict discharge consents as to the amount of suspended solids and other chemical parameters permitted. In order to comply with this, the contractor will install settling lagoons. There are usually two, sometimes three, lagoons in series, and their long retention times allow particulate matter to settle out and biological oxidation to proceed. The water may then be discharged safely to the water course affected, subject to routine checks on quality.

3.10   Protection of cultural artefacts

This is an extremely wide and contentious issue, which is largely beyond the scope of this book. In many cases there may be culturally sensitive features of sites, which should remain undisturbed. There should be satisfactory provision for use and worship if the site is of a spiritual nature, with well-defined boundaries and access points.

3.11 Protection of off-site resources

Many of the civil engineering and other activities responsible for degrading land inevitably give rise to potentially polluting materials and energies. These may include noise, dust, chemical plumes and thermal energy. Provision must be made to prevent the transmission of these off-site. In the case of noise and dust, this is often achieved by the judicial placing of top- and subsoil storage mounds. In contaminating industries care must be taken in the disposal of waste, and the treatment of gases to prevent fugitive emissions.

3.12 Conclusions

It is apparent that in order to achieve a satisfactory reclamation or restoration it is necessary to plan for it prior to any change of use likely to result in degradation. It is clear that the theory is becoming increasingly well understood and the technology is being developed in order to effect the necessary protection or maintenance of biological potential, and the reinstatement of hydrological and geological features. What we will consider in the next chapter are those sites where the planning of after-use is inadequate owing to the nature of the site or, more commonly, where no planning for restoration/reclamation was carried out in the first instance.

References

ABDUL-KAREEM, A. and MCRAE, S. (1984). The effects on topsoil of long-term storage in stockpiles. *Plant Soil*, **76**, 357–363.

BUNCE, R.G.H. and JENKINS, N.R. (1989). Land potential for habitat reconstruction in Britain. In Buckley, G.P. (ed.) *Biological habitat reconstruction*. Belhaven Press, London.

BYRNE, S. (1990). *Habitat transplantation in England: a review of the extent and nature of the practice and the technique employed.* English Nature, England Field Unit.

CAIRNS, J. JR (1993). Can microbial species with a cosmopolitan distribution become locally extinct? *Speculations Sci. Technol.*, **16**(1), 69–73.

CAREY, T. (1994). A field guide to the preservation of trees during the development process. *Living Heritage*, **11**(2), 44–45.

DOWN, G.S. and MORTON, A.J. (1989). A case study of whole woodland transplanting. In Buckley, G.P. (ed.) *Biological habitat reconstruction*. Belhaven Press, London.

HARRIS, J.A. and BIRCH, P. (1990). The effects of heavy civil engineering and stockpiling on the soil microbial community. In Howsam, P. (ed.) *Microbiology of civil engineering*. E & F. Spon, Chapman and Hall, London.

HARRIS, J.A., BIRCH, P. and SHORT, K.C. (1989). Changes in the microbial community and physico-chemical characteristics of topsoils stockpiled during opencast mining. *Soil Use Man.*, **5**, 161–168.

HELLIWELL, D.R. (1989). Soil transfer as a means of moving grassland and marshland vegetation. In Buckley, G.P. (ed.) *Biological habitat reconstruction*. Belhaven Press, London.

LIKENS, G.E. and BORMAN, F.G. (1975). An experimental approach to New England landscapes. In Hasler, A. D. (ed.) *Coupling of land and water systems*. Chapman and Hall, London.

MARRS, R.H. and GOUGH, M. W. (1989). Soil fertility – a potential problem for habitat restoration. In Buckley, G.P. (ed.) *Biological habitat reconstruction*. Belhaven Press, London.

MORGAN, J.P. and COLLICUTT, D.R. (1994). Seed stripper harvesters: efficient tools for prairie restoration. *Restoration Man. Notes*, **12**(1), 51–54.

PYWELL, R.F., WEBB, N.R. and PUTWAIN, P.D. (1994). Soil fertility and its implications for the restoration of heathland on farmland in southern Britain. *Biol. Conserv.*, **70**, 169–181.

RAMSAY, W.J.H. (1986). Bulk soil handling for quarry restorations. *Soil Use Man.*, **2**, 30–39.

SCULLION, J. (1994). *Restoring farmland after coal: the Bryngwyn Project*. British Coal Opencast Executive, Mansfield.

SPARLING, G.P., HART, P.B.S., AUGUST, J.A. and LESLIE, D.M. (1994). A comparison of soil and microbial carbon, nitrogen and phosphorous contents, and macro-aggregate stability of a soil under native forest and after clearance for pastures and plantation forest. *Biol. Fertility Soils*, **17**, 91–100.

STOCKING, M. (1995). Soil erosion and land degradation. In O'Riordan, T. (ed.) *Environmental science for environmental management*. Longman, Harlow.

WEBB, N.R. and THOMAS, J.A. (1994). Conserving insect habitats in heathland biotopes: a question of scale. In Edwards, P.J., May, R.M. and Webb, N.R. (eds) *Large-scale ecology and conservation biology*. Blackwell Scientific, Oxford.

WILSON, K. (1985). *A guide to the reclamation of mineral workings for forestry*. R&D Paper 141. Forestry Commission, Edinburgh.

Chapter 4

Former industrial sites

4.1 Introduction

Environmental legislation and regulation demand careful control of industrial waste disposal and emissions to air, land and water. Past industry did not have the degree of control we have today nor the technology to achieve it, and so in the industrial nations of the world there is a legacy of former industrial sites which may present a constraint on future development. Some of these sites may be actively polluting and a threat to humans or the wider environment unless treated. Others may be derelict or pose a threat to some future use. Yet others may have features of wildlife, archaeological or mineralogical value, derived from their industrial past but which are now thought important. Some sites may present all of these features. An indication of the nature of former industrial sites in the United Kingdom can be obtained from the survey of potentially contaminated sites in Wales (Table 4.1).

Sites can become derelict and remain so for many years, whereas for others there is pressure to redevelop as soon as they become unoccupied. Factors affecting whether a site becomes derelict or not include:

1. Location – is the site in an area where demand for land is high?
2. Size – is it of a suitable size for redevelopment?
3. Access – to vehicles, rail, port facilities and utility services.
4. Ownership.

Table 4.1 Potentially contaminated sites recorded in Wales (Welsh Office, 1988)

Site type	Number	% of total
Waste tips	279	37.4
Metal-mines	120	16.1
Iron/steel/tinplate works	106	14.2
Gasworks/coke ovens	100	13.4
Transit areas	43	5.8
Chemical works	30	4.0
Metal smelters	22	3.0
Other	45	6.1

5. Planning policies.
6. Topography.
7. Physical constraints, e.g. poor ground conditions.
8. Chemical constraints, e.g. contaminated land.
9. Cost of reuse.

Reclamation of a site may be promoted because the site is:

1. Needed for development.
2. Polluting and needs treatment to alleviate this.
3. Unsafe.
4. Unsightly.

For the most frequent types of potentially contaminated site in Wales (Table 4.1), gasworks and metal-mine sites, the pressure to reclaim gasworks is likely to arise because they are on desirable sites for development in towns, whereas metal-mine sites are in rural areas of some beauty and the pressure may arise through considerations of pollution of water courses or visual impact.

The reuse of industrial land is generally considered to be preferable to the taking of green field sites for industrial use. In order to attain the desired end-use the process of assessment, generation of design options, design, implementation and management has to be followed. To be successful, the reclamation scheme must:

1. Facilitate the desired use of the site.
2. Reduce any negative aspects of the site.
3. Enhance any positive aspects of the site.
4. Be cost-effective.
5. Not place avoidable long- or short-term constraints on the future use of the site.
6. Provide for ease of maintenance and management.

In order to achieve these aims it is crucial that the site and its relationship with the environment and its users are properly understood. Site assessment is the process of achieving an understanding which will allow an appropriate reclamation scheme to be designed and implemented. In this chapter, the principles of site assessment for former industrial sites will be discussed, followed by examples of the techniques of land reclamation, focusing particularly on the remediation of contamination. The restoration of vegetation will be dealt with in subsequent chapters.

4.2   Site assessment principles

The successful assessment of a contaminated site is crucial to the preparation of an adequate reclamation design. Site assessment usually includes:

1. A walkover survey.
2. A desk study, with preliminary investigations.
3. Detailed investigations and surveys.
4. Analysis of the information collected.

It should identify:

1. Risks presented by the site to people or the environment.
2. Constraints on the future use of the site, for example poor ground conditions, areas of contamination or steep slopes.
3. The opportunities presented by the site, for example wildlife value or the presence of structures of historical importance.
4. Structures and materials which could be put to beneficial use.

The nature of the assessments carried out is related to the future site use. Investigations of the site should be specific to the after-use concerned, for example reclamation for public open space or amenity woodland will require the investigation of soils and water for their ability to support growth of suitable plants, whereas reclamation for industrial development requires the investigation of ground conditions for foundation design and to establish the presence and extent of buried foundations. Both uses would require the investigation of the extent of contamination and its potential effect on subsequent users of the site or other targets such as groundwater.

4.2.1 Walkover surveys

Walkover surveys should take place during the desk study and at other stages of the assessment, to verify documentary information and provide information on the current status of the site. A walkover survey will yield information of the type listed below:

1. Current land-use.
2. Character of the surrounding landscape.
3. The visual impact of the site.
4. The presence of buildings, with an assessment of their structural soundness, their potential for reuse and their historic value.
5. The presence and condition of other structures, e.g. walls, culverts and bridges.
6. The nature of materials at the surface, particularly whether they impede investigative excavations at the site.
7. The presence or absence of vegetation, the nature of the vegetation and an assessment of its ecological and landscape value.
8. Deposits of waste materials.
9. The existing landforms and the constraints or opportunities they present.
10. Soil or soil substitute resources available on-site.

11. Signs of contamination, for example unusually coloured materials, odours, lack of vegetation, presence of tanks or drums which may have held, or still hold, hazardous materials.
12. Surface hydrology.
13. The presence of utilities (such as water or gas mains, electricity cables, sewers).

Specific guidance on how to assess contaminated sites during walkover surveys using sight and smell is given by the Department of the Environment (1994a).

4.2.2   Desk study

Information from the desk study will complement that of the walkover survey and include information of the following type:

1. Planning policies related to the site and the surrounding area.
2. Land ownership and other rights over land, including rights of way.
3. Current land-uses, of the site and surrounding land.
4. Current infrastructure (roads, railways, utility services).
5. Information on geology, hydrology, hydrogeology, soils and climate.
6. Positions of shafts and mine workings.
7. All former uses of the site and surrounding land.
8. Layout of plant and former process activities.
9. Waste disposal practices and licences issued.
10. Industrial archaeology, including any designation or listing of particular site features.
11. Any reports on the ecology of the site.
12. Previous site investigation reports.

4.2.3   Detailed investigations

Investigations on-site often need to be carried out by a range of specialists to address all the issues properly. It is important that a preliminary assessment of any hazards on-site is made so that appropriate precautionary action is taken for site investigation staff. Such action might include provision of protective clothing or stipulation that staff should not attend the site unaccompanied. In the United Kingdom employers have to carry out an assessment under the Control of Substances Hazardous to Health (COSHH) regulations where toxic substances may be on site, and health and safety assessments also have to be made. Typical detailed investigations carried out on a derelict site are:

1. Topographical survey.
2. Structural engineering assessment.
3. Investigation of ground conditions – physical and chemical.

4. Record and assessment of drainage conditions.
5. Location and assessment of archaeological features.
6. Record of contamination status of waters on-site.
7. Location of buried structures, assessment of their condition and likely status.
8. Ecological assessment.
9. Visual assessment.

Further details of engineering investigations may be found in Richards *et al.* (1993).

4.2.4 Analysis of information collected

All the information collected, whether from desk study, walkover or detailed investigations, should be analysed in a holistic way to determine remedial treatments in order to allow the desired use of the site. Some of the skills needed to reclaim or develop the site will be similar to those needed for uncontaminated sites, for example, structural engineering or landscape design. Others, however, such as the interpretation of chemical analyses and design of remedial treatments, will be specific to contaminated sites. In the following section some of the more important factors to be taken into account in the determining of contamination status and hence the parameters for reclamation design, will be discussed.

4.3 Assessment of contamination status

4.3.1 Sampling pattern and frequency

Contamination of the ground is associated typically with waste materials, which may have been used as fill to raise ground levels or have been disposed of on-site. Contaminants may migrate from the wastes into surrounding uncontaminated materials. The extent of such migration is dependent on the nature of the contaminants in the waste materials, on their mobility and on the nature of the surrounding ground. Contaminated ground may also arise where mobile substances, particularly liquids, have leaked from pipes or tanks, or have been spilt during on-site operations or demolition. Unlike waste materials which can be identified visually, contaminated natural ground may be visually indistinguishable from uncontaminated ground. Contaminants may also have migrated from outside the boundaries of the site under investigation.

It is impractical to excavate and analyse the whole of a potentially contaminated site to determine the concentration of contaminants present. Representative samples of the materials on-site have to be taken but there is always uncertainty as to what is representative. Decisions have to be made with regard to the number and spacing of samples taken so as to obtain information on the contamination status of a site. The degree of certainty with which these decisions can be made will be increased by basing them on information, obtained during the desk study, concerning the location of contaminating activities and waste disposal practices. In contaminated land investigations the degree of certainty is often expressed as the chance, or statistical probability, of finding a 'hot spot' of contamination; that is, a concentrated point source. The statistical probability is dependent on the sampling pattern and the frequency of sampling. Sampling patterns have been reviewed by Ferguson (1992), Richards *et al.* (1993) and Department of the Environment (1994b).

Sampling pattern   There are three basic approaches to determining the pattern of sampling:

1. **Judgemental**. Samples are taken at locations selected on the basis of prior knowledge of contaminant distribution. Such sampling is unlikely to produce samples which are representative of the site as a whole, but it is an efficient way of obtaining information on the concentration of contaminants in an area known to be heavily contaminated or the extent of migration of a contaminant from a known source.

2. **Systematic**. Sampling locations are defined by a grid system, generally a square grid (see Fig. 4.1). A grid is easy to set out on-site and is generally the method chosen where there is little prior information on the location of contamination. However, if the pattern of contamination coincides with the pattern of the grid, the samples obtained will not be representative of the site as a whole. Using a herringbone pattern rather than a simple square grid pattern reduces the risk of contamination coinciding with the grid pattern (Fig. 4.1).

3. **Random**. Mathematically determined random sampling allows statistical analysis of results to be undertaken. In its simplest form, sample points are placed randomly over the whole site. Random sampling is inefficient in that, unless a very large number of samples are taken, there may be substantial areas where no samples are taken at all (see Fig. 4.1). More sampling locations are thus required to give the same probability of locating a 'hot spot' of contamination as with systematic sampling. By using stratified sampling, that is, dividing the site into a number of areas (e.g. equal-sized squares) and placing a sampling location randomly within each area, less sampling points are needed than for simple random sampling (Fig. 4.1). An element of judgemental sampling may be introduced by varying the relative sizes of the area according to prior knowledge of the distribution contaminants across a given site.

There are various formulae proposed for deciding sampling frequency based on:

Hot spot of contamination

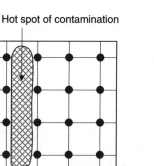

Hot spot of contamination

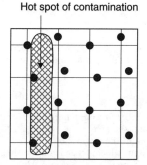

(a) Regular (square) grid pattern

(b) Herringbone pattern

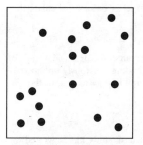

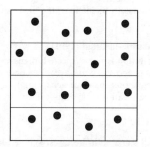

(c) Simple random pattern

(d) Stratified random pattern

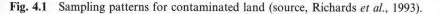

Fig. 4.1 Sampling patterns for contaminated land (source, Richards *et al.*, 1993).

1. The maximum size of a 'hot spot' of contamination which if not discovered would not cause unmanageable problems later in the development of the site.
2. The degree of certainty of finding such a hot spot (Table 4.2).

Similarly both British and Dutch codes of practice for investigating contaminated land make tentative recommendations for the number of samples which should be taken at any one location (British Standards Institution, 1988; Nederlands Normalisatie-Instituut, 1991). In practice, the most effective procedure is usually a regular sampling pattern, modified by desk study considerations and the experience of the investigator. Frequency and number of samples are similarly determined; however, the formulae for determining frequency do provide a guide to what, on statistical grounds, would be a sensible number of sampling points. The iterative nature of the process must be stressed; an experienced investigator will continually reassess the information to hand from desk study, previous trial pits and investigations; modifying the approach through the investigation.

In some situations, the investigation of contamination is carried out in more than one stage, for example, the initial ground investigation may be followed

Table 4.2   Formulae for determining sampling frequency for contaminated land

Ferguson (1992)

In the herringbone sampling pattern the number of sampling locations needed to ensure a 95% probability of hitting a target 'hot spot' is given by the equation:

$$N = \frac{kA}{a}$$

where: N is the number of sampling points
A is the total site area
a is the area of the target 'hot spot'
k is a constant which depends on the shape of the target as follows:
circular target, $k = 1.08$;
plume-shaped target, $k = 1.25$;
elliptical target, $k = 1.80$.
An assessment thus has to be made of the likely shape of the target in order to calculate the sampling frequency.

Nederlands Normalisatie-Instituut (1991)

For sites where pollution is not suspected intensive sampling and analysis for a wide range of possible pollutants is necessary to have a reasonable chance of finding previously unknown pollution. Sampling should be done according to a systematic sampling pattern, with the number of sampling locations for near-surface samples, n, given by:

$$n = 10 + 10A$$

where A is the total site area in hectares. Where the preliminary desk study has suggested that pollution is present but that its distribution is homogeneous the number of sampling locations is given by:

$$n = 5 + A$$

Where contamination is thought to arise from known point sources, four sampling locations per point source are recommended, with at least one groundwater observation well per point source. Where the locations of the point sources are not known, the recommended number of sampling locations is given by:

$$n = 4 + \frac{A}{a}$$

where a is the estimated area of contamination in hectares.

---

by the preliminary design of reclamation proposals and then by further ground investigations to clarify reclamation options. Similarly, where it is proposed that contaminated materials be removed, further investigation will be needed to define the quantities and pre-use characteristics of the materials involved. However, first stage surveys involving very low sampling frequencies, provide little reliable information on the overall contamination status of a site.

A staged site investigation will have the advantage of focusing a contamination assessment on a particular area or areas of contaminated ground, and is likely to provide more accurate information about critical areas of a site.

4.3.2 *Analysis of samples*

Analysis procedures and details of individual analytical methods are outside of the scope of this book. A number of countries have compiled analytical methods for contaminated land (United States Environmental Protection Agency, 1992; Canadian Council of Ministers of the Environment, 1993) and in the United Kingdom methods are recommended by the Department of the Environment for those contaminants for which standards have been suggested. Similarly, the National Rivers Authority recommends particular leaching tests for determining the potential of contaminated ground to pollute controlled waters (National Rivers Authority, 1994). Whatever the analysis method or the determination there are 'ground rules' which should be adopted when dealing with contaminated land:

1. Analysis of samples is often the most expensive part of a contaminated land investigation – care has to be taken in deciding which samples to analyse and for which constituents.
2. Samples should be kept, under conditions where chemical changes will not occur, until reclamation proposals are finalized, in case further analysis is required.
3. The analysis suite should be chosen on the basis of:
    (a) substances thought to be present on the basis of the past uses of the site;
    (b) substances which are thought likely to cause a hazard given the proposed use of the site;
    (c) observations of the materials present made during the site investigation.
4. Regulations in force or standards set may result in pressure to analyse for other determinands. These analyses should be in addition to those based on the criteria above.
5. Screening analysis which indicates the presence of a group of substances but not the concentration of individual compounds can be used to gain maximum information for minimum analysis expenditure.
6. The person specifying the analysis suite and interpreting the results should have a good understanding of the methods used, and their advantages and limitations.
7. Reporting of results should include information about the methods used so that correct interpretations can be made.
8. Variation in results is much more likely to be due to unrepresentative sampling than analytical errors. If the analysis budget is limited then it is much better to analyse a large number of samples using a moderately accurate method, rather than a few samples using a very accurate but costly method.
9. For groundwater samples very low concentrations may be of interest. Therefore for these samples the limit of detection of the analytical method

employed should be lower than the lowest concentration which could be of concern.
10. Quality control should be rigorous.

4.3.3 Interpretation of results

The correct interpretation of analytical results is critical to the successful design of a reclamation scheme. The analytical results have to be considered in conjunction with the information on the past use of the site and all of the information gained from the site investigation, including specialist reports and consultations. The aim of the interpretation of the results is to provide information to allow design of reclamation to proceed. A principal step in fulfilling this objective is to decide what concentrations of contaminants are regarded as significant.

We will address three different approaches to this problem:

1. Comparison with 'natural' concentrations in unpolluted soils.
2. Comparison with concentrations in soil surrounding the site.
3. Consideration of the risks associated with the presence of the substances at different concentrations.

Whilst any increase in soil concentrations of hazardous substances above 'natural' levels may be considered to be undesirable, in practice comparison of soils from heavily industrialized urban areas, where the majority of land may contain elevated concentrations of potentially hazardous substances, with concentrations found in soils of unpolluted rural situations may lead to unjustified conclusions and an unnecessary level of clean-up.

Comparison with local background soil concentrations is a more practical approach, which allows areas containing high concentrations of contaminants to be identified and resources directed at treatment of these, to give a gradual improvement in the contamination status of an area over time. This approach of developing standards specific to a local area is also necessary where the background concentrations are naturally high, for example in the south-west of England where arsenic concentrations in soils may naturally be as high as several hundred mg/kg. Care has to be taken, however, that locally high background concentrations are investigated in their own right to ensure that the risks associated with them are tolerable.

The third type of approach considers the risks associated with the presence of the substance in question, and attempts to define concentration values below which the risk is negligible and above which the risk is unacceptable. Evaluation of risk implies consideration of the system:

$$\text{Contaminant} \quad \overset{\text{Route}}{\Longrightarrow} \quad \text{Target}$$

In order for a contaminant to reach a target in a dose at which it will cause damage, all three components, pollutant, route and target, need to be present. If any one component is removed then there may not be a hazard.

In this model the target may be people, animals, plants, water resources or building materials, but can also be considered more widely such as financial or commercial assets. Routes by which targets may be exposed to contaminants are outlined in Table 4.3. The degree to which routes of exposure operate depends, amongst other consideration, on the use of the site. Therefore, the risk assessment approach frequently takes into consideration the intended use. Assessment of risks to humans and animals from toxic substances in soils involves determining acceptable dose levels by consideration of exposure routes.

Site-by-site risk assessment and comparison with nationally set generic values is an approach which many nations have adopted. The United Kingdom has always had a 'use-based' approach to the assessment and clean-up of contaminated land. This approach results in clean-up of sites to a standard which allows its next use, rather than one which necessarily allows a range of uses of different sensitivities. The approach does, of course, allow for clean-up to prevent contaminants reaching non-use-based targets such as groundwater. Rigorous risk assessment has not, however, often been carried out, partly because it can be a laborious (and expensive) process but more often because not all the data on pathways and toxicities have been available. Even when such data are available a high degree of judgement has to be used in selecting data and accepting conclusions. Nevertheless, interpretations which

Table 4.3 Principal hazard pathways and targets on contaminated land

Target	Pathway	Secondary targets
Human health	Ingestion, inhalation, dermal contact, fire/explosion	
Flora and fauna	Animals: ingestion, inhalation, dermal contact Plants: absorption through roots, absorption through aerial tissues, coating of aerial parts and blocking of stomata with particulate matter	Movement through food chain to other plants and animals (including domestic animals) and humans. Loss of habitat and species. Loss of genetic diversity – tolerant species/ecotypes survive
Air quality	Volatilization of contaminants, increase in particulate loading	Humans, flora and fauna, soil and water quality
Soil quality	Migration of contaminants in soil and groundwater, deposition of particulate matter	Wildlife, crops and domestic animals, surface and groundwater
Surface and groundwater	Direct contact resulting in solution of contaminants, suspension of contaminated soil particles. Aerial deposition on water courses	Humans, wildlife, crops and domestic animals and soils
Construction materials	Direct contact, migration in groundwater	Humans
Buildings	Fire/explosion, heave, subsidence, migration of gases and liquids	Humans

rely on risk assessment principles are likely to be the most robust and provide the best reclamation designs, and some detail of the risk assessment process will be given here.)

Figure 4.2 shows how risk assessment relates to the whole risk management process and interacts with investigation, reclamation design, management and monitoring of sites. Figure 4.3 indicates how the identification and assessment of the hazard caused by an individual contaminant might be approached. Hazard assessment involves considering the pollutant/route/target model, and Fig. 4.3 refers to 'background' and 'threshold' concentrations. These are guidance thresholds usually set at a national level and with which concentrations of contaminants found on-site may be compared. The thresholds are usually derived from toxicological data and there may be different thresholds for different end-uses. In the United Kingdom, these levels are set by the Interdepartmental Committee on the Redevelopment of Contaminated Land (ICRCL), a committee of the Department of the Environment. It is worth noting that such levels are for guidance only. The possibility that a contaminant may be present in concentrations below background level and constitute a hazard should always be considered. More often contaminants may be found in concentrations above both background and threshold levels but may not represent a hazard as there is no pathway to reach the target (for example, high levels of lead at a lead mine will not necessarily represent a hazard to a hill walker who walks through the site). The ICRCL levels adopt two trigger levels: the lower is termed the threshold trigger concentration below which the site is considered uncontaminated for the particular use proposed; the upper is termed the action level above which remedial action is necessary (ICRCL, 1987). Between threshold and action trigger levels is an area where 'professional judgement' should be exercised as to whether remedial treatment is necessary.

Careful selection of standards is necessary to ensure that they are appropriate for the situation, for example standards set for residential use should not be applied to a site intended for industrial use, and that note is taken of any incorporated assumptions. (For example, are the standards mandatory in the country of origin? What is the policy basis for setting the standards?)

For many situations a qualitative risk assessment based on an appropriate hazard assessment will be sufficient to determine reclamation design parameters. Quantitative risk assessments may be needed when the frequency and level of exposure to contaminants causing significant effects are high, public perceptions of risk are high or local background concentrations are high relative to threshold standards, and there is a risk of the site contributing to the local environmental burden. The procedure for quantitative risk assessment is outlined in Table 4.4. The results of such an assessment are typically compared with reference exposure values in order for judgements to be made with respect to treatment measures needed. Both qualitative and quantitative assessments lead to the determination of parameters for reclamation design. The Welsh Development Agency introduces, in its manual on the remediation of

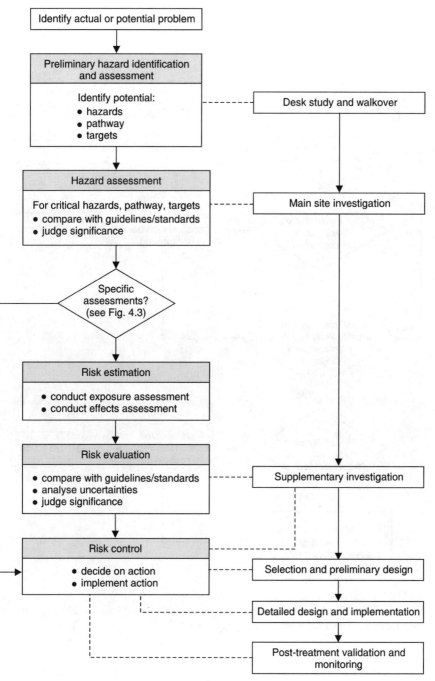

Fig. 4.2 Interrelationships between risk assessment and the investigation, design, management and monitoring process (source, Welsh Development Agency, 1993).

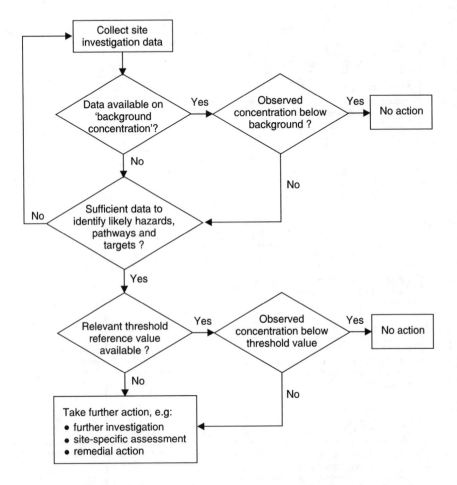

Fig. 4.3 Identification and assessment of the hazard caused by an individual contaminant (source, Welsh Development Agency, 1993).

Table 4.4   Outline of quantitative risk assessment procedures (Welsh Development Agency, 1993)

---

1. **Exposure assessment** – quantification of the environmental transport and fate of contaminants, including:

- the concentration of contaminants at the source, at points along the travel pathway and at the point of exposure (e.g. ingestion by the target)
- the rate at which contaminants move along the pathway
- the chemical form and physical properties of the contaminants
- the characteristics of the host medium (soils, rock, groundwater, etc.) and the extent to which environmental factors such as dispersion, dilution, degradation, adsorption, etc. modify contaminant concentrations as they move along travel pathways
- how much, how often and over what period exposure at the point of the target is likely to take place
- the characteristics of the exposure route (e.g. ingestion, inhalation, direct contact) that determine how much of the contaminant is taken in by the target

2. **Effects assessment** – determination of the effect (e.g. toxicological, carcinogenic, mutagenic, corrosive) of the hazard on the target under the conditions of exposure defined by 1 above. Effect assessments involve a consideration of:

- the dose/response relationship for the contaminant being assessed (in particular the nature of the response at or below the no observable effect level (NOEL))
- data on the biological mechanisms regulating responses to different types of substances
- the characteristics of the target (e.g. gender, age, general health status, species composition, physical properties of the building fabric) since these influence the dose/response relationship
- the limitations of toxicological and other reference data to adequately describe effects produced under environmental exposure conditions (e.g. the extent to which animal experiments can be safely extrapolated to the human situation)

3. **Assumptions/uncertainties** – in practice risk estimation can be subject to uncertainty owing to a lack of sufficient data on, for example:

- the chemical form, behaviour and concentrations of contaminants reaching the target
- the nature of the exposure (how much, how often and over what period)
- the effect of the exposure on the target (especially at low dose levels and in relation to carcinogenic hazards)

Uncertainties are typically addressed using assumptions. Worst case assumptions are often employed to build adequate margins of safety into the assessment, e.g.

- the contaminant is present in its most toxic/available form
- the highest observed concentrations are typical of the contaminated area as a whole
- minimal dilution, dispersion or degradation, etc. of the contaminant occurs along the travel pathway
- all the material is biologically available, reaches the target organ and has a full effect
- the maximally exposed and most sensitive individual or species is representative of the population at risk

---

contaminated land (Welsh Development Agency, 1993), the concept of contaminated land remediation objectives (CROs). This is an effective concept allowing focus on quantitative values for clean-up design, management and monitoring. CROs can be derived from threshold values (e.g. ICRCL limits), threshold values modified to take site-specific values into account or values derived from site-specific risk assessments. Different CROs may be derived for different parts of the same site depending on site use and targets at risk.

4.4   Reclamation techniques

4.4.1   The range of techniques available

The aim of reclamation is to return land to beneficial use or to prevent contaminants on land from causing damage. Former industrial sites will have a range of constraints identified in the site assessment. These constraints may include engineering constraints, such as poor or unstable ground and unstable or undesirable structures; visual constraints, such as unacceptable landforms or unsightly fly-tipping; and chemical constraints, such as contaminated land and water. Reclamation is the activity that treats land to overcome these constraints. As all sites are different in the opportunities and constraints they offer and the uses to which they will be put, no two reclamation schemes are the same. However, there are generic techniques which can be used in isolation or in combination to treat sites.

Cost and the regulatory framework are the principal determinants of the choice of reclamation method and these factors account for the differences in emphasis between reclamation methods used in different countries. So, for example, in the United Kingdom there is more landfilling than in Germany and the Netherlands, because landfilling is cheaper and the controls on the landfilling of certain wastes not as strict. Stricter landfill controls in the United Kingdom do, however, mean that other techniques of dealing with wastes are becoming used more often.

It should be a primary objective of those responsible for the reclamation of sites containing toxic materials to ensure, not only that these materials are dealt with in such a way as to reduce risk, but also to implement schemes that aim to conserve uncontaminated natural materials as far as possible. The treatment of contaminated land should therefore:

1. Reduce the amount of toxic materials being dealt with by confinement in landfill.
2. Increase the separation of contaminants from the matrix which they are polluting, so reducing the volumes of waste generated, and conserving soil and water.

The reclamation methods available divide broadly into:

1. Traditional methods of isolation, covering or removal of contaminated materials.
2. Decontamination techniques which remove contaminants from contaminated land and water.

Isolation and removal techniques practised are presented in Table 4.5. Decontamination treatments can be broadly divided into three categories for soil and two for water, as illustrated in Fig. 4.4. The application of such treatments to various contaminants is presented in Fig. 4.5.

Table 4.5 Options for the isolation and/or removal of contaminated ground

Method	Characteristics
Isolation	
Capping	Impermeable layer covering the contaminated ground to reduce rainfall infiltration
Covering	Similar to capping but may not be impermeable or contain drainage layers to remove infiltrated rainfall
Vertical barriers	Impermeable, vertical, subterranean barriers to minimize migration of contamination and movement of contaminated groundwater
Diversion trenches	Drainage systems to intercept water and/or pollutants from contaminated ground
Break layers	Layers of single-sized stone above saturated zone of contaminations to prevent upward movement of contaminants by capillary action
Horizontal barriers	Impermeable barriers beneath waste or contaminated ground to reduce downward movement of contaminants
Removal	
Excavation	Physical excavation of contaminated ground and removal to a suitable landfill or treatment plant
Total containment	Excavation of material and replacement in a purpose-built impervious cell

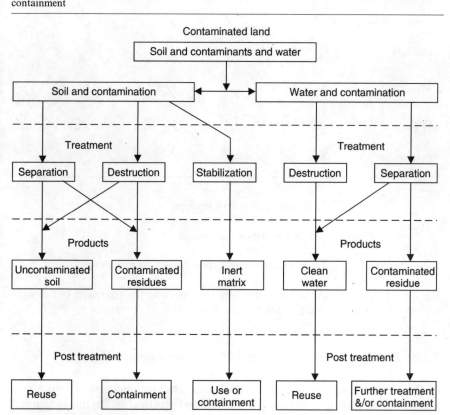

Fig. 4.4 Application of treatment technologies to contaminated materials and the implications for production formation and further treatment (source, Richards *et al.*, 1993).

Target compounds	Treatment technology										
	Vapour extraction	*In situ* bioremediation	*Ex situ* bioremediation	Soil washing	Containment	Stabilization/ solidification	Thermal treatment	Vitrification	Solvent extraction	Pump and treat	Leaching
Metals (fines and soluble)	✗	✗	✗	✓	✓	✓	○	✓	✗	○	✓
Metals (larger particulate)	✗	✗	✗	✓	✓	✓	✗	✓	✗	✗	✗
Volatile organic compounds	✓	✓	○	✗	✗	✗	○	○	✓	✓	✗
Semi-volatile organic compounds	○	✓	✓	✓	✗	✗	✓	○	✓	✓	✗
Halogenated organics	✓	○	✓	✓	✗	✗	✓	○	✓	✓	✗
Oil hydrocarbons	○	✓	✓	✓	✗	✗	✓	✗	✓	✓	✗
Coal tars	✗	○	○	✓	✓	✓	✓	✗	✓	○	✗
Asbestos	✗	✗	✗	✗	✓	✓	✗	✓	✗	✗	✗
Coal	✗	✗	✗	✓	✓	✗	✗	✗	✗	✗	✗
Dioxins	✗	✗	✗	✓	✓	✓	✓	✓	✓	✗	✗

✗ : Inappropriate in most cases

✓ : Appropriate in many cases

○ : Of some potential under certain circumstances

Fig. 4.5 Target soil and groundwater contaminants and the treatment technologies appropriate for these contaminants (source, Richards *et al.*, 1993).

4.4.2 Isolation, covering or removal of contaminated materials

In the United Kingdom the most commonly used technique for dealing with contaminated soils is covering. Excavation and removal to landfill is also frequently used. Although neither technique degrades or destroys the contaminants and is therefore not as environmentally 'fail-safe' as techniques which do, the effectiveness and cheaper cost of covering and excavation mean that they will continue to be much-used techniques. The requirements and application characteristics of isolation, covering or removal techniques are summarized in Table 4.6.

Cover systems can be quite complex (Fig. 4.6). An indication of how cover systems have evolved can be gained from a consideration of the systems which

Table 4.6 Applications and limitations of isolation, covering or removal techniques for contaminated land

Method	Application	Limitations
Capping and covering	• Isolation of contaminants from potential targets • Prevention of exposure to transport media, e.g. water and wind • Control of gas, leachate and capillary water • Ground improvement for construction • Providing a substrate for vegetation	• Conflicts between differrent aims may be difficult to overcome, e.g. an impermeable cap may need to be breached to allow piled foundations • May lack permanency • Susceptible to damage • Requires specialist input in design and construction to ensure meets performance standards • Ineffective against lateral groundwater movements
Diversion trenches	• Cut-off drainage around contaminated areas • Temporary measures during excavation and to facilitate removal of contaminants during pump and treat operation or to lower groundwater table to facilitate other measures	• Limited applicability to impermeable soils • Need for specialist input • Requires long-term monitoring and maintenance
In-ground barriers	• Minimize lateral or vertical movement of contaminants, gases and water • Vertical barriers are of three principal types: displacement barriers, e.g. sheet piling; excavated barriers, e.g. slurry trenches, membranes; injection systems, e.g. jet grouting • Horizontal barriers are used where a contaminant source is underlain by permeable materials and are usually installed using injection/ grouting methods	• Specialist input needed • Difficult to install in heterogeneous ground conditions • Requires long-term maintenance and monitoring • Vertical membrane installation can be difficult and if not properly installed can lead to costly remediation measures • Sheet piles are costly to install and not impermeable
Excavation	• Used to remove materials from site for final disposal or treatment • Disposal may be on- or off-site with or without treatment • Both contamination and poor ground conditions are remediated at the same time by excavation	• Complete removal of contaminants may not be achieved because of proximity of existing structures or utility services or because contamination extends too deep for excavation • Does not destroy contaminants • Long-term maintenance and monitoring of disposal areas required

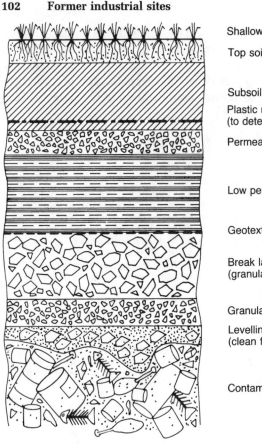

Shallow rooted vegetation

Top soil

Subsoil

Plastic mesh
(to deter deep excavation)

Permeable drainage layer

Low permeability clay

Geotextile layer

Break layer
(granular fill)

Granular separation

Levelling layer
(clean fill)

Contaminated wastes

Fig. 4.6 A cover system for contaminated land incorporating a clay cap and break layer (source, Richards *et al.*, 1993).

have been in use in the United Kingdom for metalliferous mine sites (Fig. 4.7). Cover systems are effective and economic, particularly where large areas of ground need to be reclaimed. Effective cover systems can be introduced as part of necessary ground improvement, such as the use of clean fill material into which utility services may be laid or buildings founded. Buildings and hard-standing areas such as car-parks can be themselves effective covers where water drains away through the surface drainage system without contact with under-lying contaminants. In all cover systems the effect of underlying contaminants on the integrity of the system must be gauged as part of the site assessment procedure and any necessary protective measures taken. Although a car-park may be an effective cover, for example, tarmacadam can become softened in contact with some hydrocarbons and would need isolating from these in the ground. Similarly, the surface water drainage system, and materials such as plastics and concrete associated with other utility services, may need protection from contaminants remaining in the ground beneath a cover.

Vertical barriers and excavation, to prevent lateral migration of contaminants or to remove the most contaminated materials, may be used in a reclamation scheme where covering is the principal method being used. Vertical barriers, in combination with a cut-off drainage system are frequently used in this way (Fig. 4.8). Horizontal barriers are less frequently used in the United Kingdom.

Excavation may be used in conjunction with containment to contain small volumes of contaminated material on-site. The containment cell is likely to be constructed of clay or bonded geotextiles with specially constructed drainage and monitoring points on the periphery of the contained material. In many instances such a containment cell will become a 'sterilized' area on which development cannot take place, and which will need long-term monitoring and maintenance.

4.4.3 Decontamination techniques

Decontamination techniques can be broadly divided into thermal, separation, solidification/stabilization, chemical and biological techniques. Decontamination may be carried out *in situ*, that is, the decontamination process takes place in the ground, or *ex situ* where decontamination takes place in excavated material. *Ex situ* techniques may be implemented on- or off-site.

Thermal treatment   There are three broad categories of thermal treatment: incineration, thermal desorption and vitrification. Incineration is a destructive technique which destroys both the contaminant and the matrix in which it is contained. So, for contaminated soils treated by incineration soil structure and function are lost, as well as the contaminant. Although mobile incinerators are available, incineration usually takes place off-site, often in large incinerators, designed to handle the more toxic waste materials. Such a plant can have a throughput of 100 000 t/year. Incinerators produce waste in the form of emissions and ash. Improvements in the design of incinerators to minimize the effect of these wastes on the environment is causing incineration to become a costly process. Some of the more common incineration methods are described in Table 4.7.

Thermal desorption is the process of removing volatile and semi-volatile compounds from contaminated materials. The processes used are varied and some have similarities with incineration techniques. As in incineration, desorbed compounds have to be collected before exhaust gases are emitted to the atmosphere. Desorption methods are described in Table 4.8.

Vitrification processes involve the mixing of silica particles or other vitrifying materials such as sand or fly ash with contaminated waste and then applying intense heat. The heat drives off the volatile compounds and causes the solid

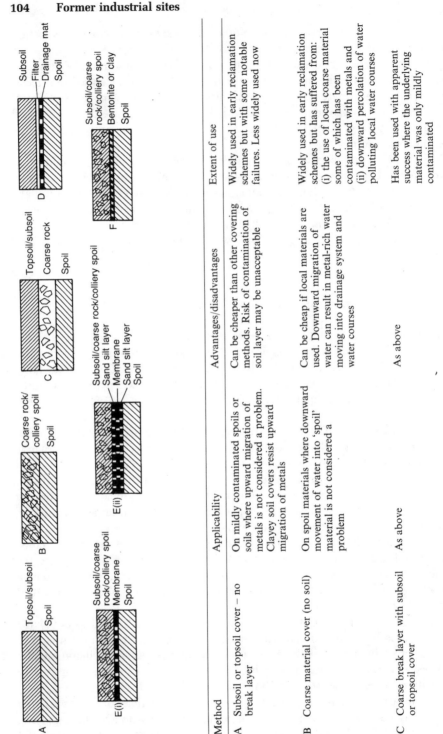

Method	Applicability	Advantages/disadvantages	Extent of use
A Subsoil or topsoil cover – no break layer	On mildly contaminated spoils or soils where upward migration of metals is not considered a problem. Clayey soil covers resist upward migration of metals	Can be cheaper than other covering methods. Risk of contamination of soil layer may be unacceptable	Widely used in early reclamation schemes but with some notable failures. Less widely used now
B Coarse material cover (no soil)	On spoil materials where downward movement of water into 'spoil' material is not considered a problem	Can be cheap if local materials are used. Downward migration of water can result in metal-rich water moving into drainage system and water courses	Widely used in early reclamation schemes but has suffered from: (i) the use of local coarse material some of which has been contaminated with metals and (ii) downward percolation of water polluting local water courses
C Coarse break layer with subsoil or topsoil cover	As above	As above	Has been used with apparent success where the underlying material was only mildly contaminated

D	Drainage mat used as break layer with cover of subsoil	As above	Synthetic layers can be cheaper than stone where materials have to be imported. Downward migration of water can result in metal-rich water moving into drainage system and water courses	In limited use
Ei	Polyethylene	On all spoil materials though on coarse angular materials a bed of sand or silt may have to be used to prevent puncturing	Prevents upward and downward movement of water and contaminants. Allows the use of any capping material. Membrane easily punctured. Cannot be placed on slopes over 1 in 6. May suffer in drought years if capping/soil layer is shallow	Used in recent years in schemes in Wales where threat of pollution of water coarses was an issue
Eii	Membrane with cover			
F	Bentonite or other clay seal with cover	On all spoil material	As above. Can be used on steep slopes. Can be expensive	Used in schemes in Wales where threat of pollution of water courses was an issue and slopes steeper than 1 in 6 have to be capped

Fig. 4.7 Cover systems used in the United Kingdom in metalliferous mine reclamation schemes (source, Department of the Environment, 1994c).

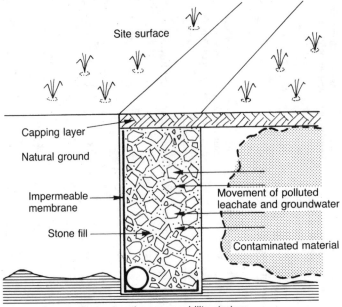

Fig. 4.8 Vertical barrier and cut-off drainage in use with a simple capping system (source, Richards *et al.*, 1993).

Table 4.7 Incineration methods

Method	Description
Rotary kiln	Refractory-lined inclined tube which is slowly rotated whilst the material within is heated. The rotation ensures that the heating is done under oxidizing conditions by pipes within the kiln or through the shell of the vessel. Retention times are different, dependent on the waste being treated, and are affected by both kiln angle of inclination and speed of rotation. A variety of wastes can be treated and the kilns are capable of a high throughput, but some wastes, e.g. clays, can form clods during treatment. The exhaust gases contain the volatile fractions of the waste undergoing treatment (e.g. contaminants and water) and these are subject to secondary treatment (gas scrubbing and particulate collection) before being emitted to the atmosphere.
Infrared	Infrared lamps are used to irradiate contaminated materials passing under them on a conveyor belt. The furnaces are run under oxidizing conditions and the exhaust gases have to undergo secondary treatment similar to that in rotary furnaces before emission to the atmosphere. The systems can run on transportable units.
Fluidized bed	Fluidized bed systems maintain a bed of fluidized solids fuelled as necessary to maintain high temperatures. The retention time is short and any incompletely combusted particles are returned to the bed. These beds have been used in a limited way to treat contaminated soils, but throughputs of 450 t/h have been achieved for waste tar from coking plant operations.

Table 4.8   Desorption methods

Method	Description
Ex situ thermal desorption	Similar to rotary kilns but acting at a lower temperature these systems can be directly or indirectly heated. Desorption takes place at temperatures as low as 320°C for volatile organic compounds (VOCs) or higher for semi-volatile organic compounds.
In situ stripping	Involves injecting steam or hot air into the ground through boreholes or drill stems. Contaminated vapours are recovered through recovery boreholes in the case of injection boreholes or via a 'tent' over the treatment area maintained at reduced pressure in the case of injection through drill stems. Vapours and gases are condensed or stripped to separate contaminants. These methods are most suited to homogeneous permeable soils.

material to vitrify, forming a glass-like compound. The vitrified material is chemically inert and can be used in construction. As with other methods, the exhaust gases have to be treated before emission to the atmosphere.

Thermal treatments that have not been demonstrated other than in experimental or pilot projects are radio frequency or electric heating of soils to volatilize contaminants and *in situ* vitrification using graphite electrodes.

Thermal treatments are an effective way of destroying the contaminants in soil but their effectiveness is dependent on grinding the feedstock to reduce variability, and careful control of operational parameters such as temperature and residence time. Disadvantages are that thermal treatments are very energy-intensive, exhaust gases have to be cleaned and soil materials are destroyed or damaged.

Separation techniques   Techniques for separating contaminants from soil particles or groundwater, or more contaminated particles from less contaminated particles, include vacuum (or vapour) extraction, soil washing, solvent extraction, electrokinetics, and leach, pump and treat. These techniques are based on different principles which are summarized in Table 4.9.

Separation techniques are becoming widely used and are particularly effective at reducing concentrations of contaminants to an acceptable level rather than complete removal. Whatever the separation method, removal of the first, say, 80 per cent of contaminant is likely to be easier than removal of the last 20 per cent; for example, with vacuum extraction the least volatile compounds will be the last to be removed and by definition the most recalcitrant. Similarly, in soil washing the particles containing the lowest concentrations of contaminant will be the last to be removed and are the least amenable to such removal. A schematic representation of a vacuum extraction is shown in Fig. 4.9. Integrated systems where wastes from one process are treated by another offer much more potential for the removal of these recalcitrant fractions.

Table 4.9   Separation techniques

Technique	Characteristics and application	Limitations
Vacuum extraction	*In situ* technique for the removal of VOCs from contaminated land by means of vacuum pumps or fans applied to extraction wells. The reduced vapour pressure resulting in the wells causes air and VOCs to flow out through the extraction wells and also causes volatilization of further contaminants. The flow of air through the system can also cause enhanced biological degradation of organic contaminants and has led to a variation of the technique known as bioventing which relies on these biodegradation processes. If biodegradation takes place, compounds of low volatility may be degraded. Extraction of VOCs is often accompanied by simultaneous but separate extraction of contaminated groundwater, a process known as dual vacuum extraction. A typical extraction system is shown in Fig. 4.9.	The principal technical limitations to vacuum extraction are soil permeability and contaminant volatility. Best results are obtained in homogeneous permeable soils and an understanding of contamination profiles and soil permeabilities at a site are crucial to efficient system operation. Extracted vapours may be emitted direct to the atmosphere or need prior treatment. Permissions for such emissions may be difficult to obtain in some localities.
Soil washing	Soil washing is a general term applied to *ex situ* techniques which use water-based processes for the separation of contaminants from soils. The techniques are largely derived from the mineral processing industry and are applicable to both inorganic and organic substances. The principal processes involved are: particle size separation where contaminants are associated with a particular size range; extraction of contaminants in wash water; suspension of contaminants in wash water. Chemicals to alter the surface properties of particles and enhanced gravity technology may be used to obtain better separation. Soil washing may be used in conjunction with other methods of treatment such as leaching and is often designed in modular fashion to suit a particular site's needs.	Silt and clay soils are less amenable to treatment than sandy and gravelly ones. The plant also needs space on-site and a source of power and water. Spent wash water will need disposal of and even though filter-pressing and recycling of water will limit the quantity to be disposed of this may cause difficulties at some sites.
Solvent extraction	Solvent extraction removes contaminants from the soil matrix into a liquid medium in a similar way to soil washing but uses solvents to selectively remove contaminants rather than water. The technology is applied principally to organic contaminants and has been used for contaminants such as PCBs and petroleum hydrocarbons. Solvents are usually recycled with separation of solvent and contaminant being carried out by steam stripping or heat or pressure changes. Liquefied gases such as propane and butane can be used as solvents with separation being achieved by depressurizing. Solvent extraction techniques can be used in conjunction with other decontamination techniques.	Concentrated wastes need disposal and will contain traces of the solvents themselves some of which can be toxic. The efficiency of the processes can be affected by organically bound metals and detergents present in the media to be extracted. The techniques are of limited applicability to soils heavily contaminated with amenable contaminants. The technique has not been widely used in Europe.

Table 4.9 (contd.)

Technique	Characteristics and application	Limitations
Electrokinetics	Electrokinetics involves establishing an electric field in a contaminated zone by installation of electrodes. The electric field forces migration of ions towards cathode and anode around which circulate acids or alkalis which capture the migrated ions. The method is applicable to fine-grained as well as coarse-grained soils and can be used in *in situ* and *ex situ* applications.	Effective only against electrically charged ions and has had limited field use.
Leach, pump and treat	Leach, pump and treat techniques are variations on soil washing and solvent extraction and are carried out *in situ*. They involve pumping water out of a contaminated zone and subsequent treatment of the water. Dual pumping systems may be used where uncontaminated groundwater is separated from contaminated water (Fig. 4.10). Leaching techniques which involve the introduction and directed flow through contaminated materials of a leaching solution may use sprinklers, pipes, trenches or boreholes to introduce the leach solution. Leach liquids may include water, acids and surfactants and are usually followed by flushing with clean water if water was not the leach liquid.	Extracted water has to be treated and it is often difficult to predict the quantity and quality of water to be treated beforehand. Best suited to homogeneous and permeable soils. Leaching methods are only suitable for use above the water table.

Solidification/stabilization Solidification is the fixing of contaminants in a resistant solid matrix. Stabilization is the treatment of contaminants to make them less available or toxic to targets. Solidification most commonly uses concrete-based materials or thermoplastics as the cementing medium, but silicates and lime are also used. Pre-treatment of non-solid materials may be necessary to separate the contaminant into a solid phase by, for example, the use of clay minerals. The process of solidification usually involves mixing a slurry of contaminated material and solidifying medium, and then solidification in moulds. Stabilization by chemical means is less well used, but the most promising methods use oxidizing agents. Although chemical stabilization would appear a promising technique, its specificity to particular contaminants and the heterogeneous nature of most contaminated soils means that its use is limited. Solidification techniques, particularly those using cement-based methods, are considered acceptable in the United States although they have been less applied in Europe.

Principal limitations are that, because of the antagonistic effect of some contaminants on the solidifying medium, a greater volume of material may result after solidification and that disposal of the solidified material has to be achieved. Heat and gases may be generated during treatment, and these have to be controlled. Solidified materials may have greater structural properties than

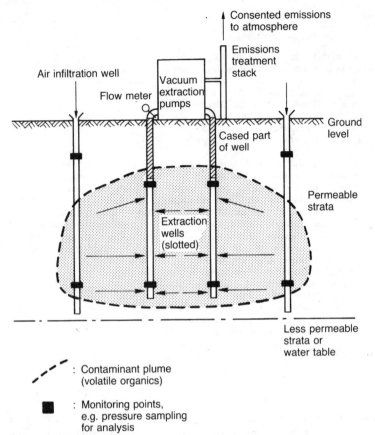

Fig. 4.9   Schematic vacuum extraction system (source, Richards *et al.*, 1993).

the original material and be disposed of on-site, increasing the construction properties of the site; however, the long-term performance of some of these solidifying agents has not been established.

Chemical methods   These use chemical processes to destroy the contaminants in a soil or to make them less toxic. An example is the use of ozone to destroy hydrocarbon contaminants. This technique has been widely investigated but is likely to be expensive, because of the energy costs of producing ozone and the fact that ozone will be consuming oxidizing, non-contaminating, organic compounds such as soil organic matter. Dechlorination is another technique which has received limited use. This technique is intended to dechlorinate contaminating halogenated hydrocarbons such as PCBs and dioxins. Dechlorination is carried out using reagents such as potassium hydroxide and polyethylene glycol, but can be adversely affected by high humic and clay contents of soils.

It is unlikely that chemical methods will become widely used in decontaminating contaminated soils because of their specificity and potential for interference by other compounds. They may find limited use where soils are homogeneous and contaminated by few substances.

Biological methods   Biological treatment uses natural biological degradation processes to degrade organic compounds; the most developed techniques making use of micro-organisms. The aim of biological treatment is to optimize those factors that stimulate biodegradation. Critical factors are: temperature, water, pH, nutrient supply, presence of appropriate microbial population and absence of microbial inhibitors. The compounds most amenable to degradation are those that have a degree of aqueous solubility, so compounds such as the heavy fractions of coal tar which have negligible solubility are resistant to biological degradation, whereas compounds such as light mineral oils which have some aqueous solubility are amenable to biological treatment.

In situ biological treatment is achieved through the circulation of groundwater dosed with amendments to stimulate biological activity. Contaminated areas may have to be 'batched' for treatment by the use of vertical barriers or boreholes. A schematic *in situ* treatment system is shown in Fig. 4.11.

If ground conditions are such that *in situ* biological treatment is not possible, for example if soil permeability is low, then excavation of material and treatment in beds is an alternative. The beds or heaps are aerated through tilling or incorporation of pipework; similarly, watering may be achieved by sprinklers or pipes. Drainage to control leachate, and collection and treatment of volatile gases is often necessary. Off-site treatment is quite popular in some countries because it allows permanent facilities for treatment of leachates and collected volatile compounds to be installed. Off-site treatment also has the advantage that commencement of other site works do not have to wait until the soil has been treated by biological means which can take some months. Excavated soil may be returned to the same site after treatment or sent to another site or sent to landfill. There has also been an increase in the use of reed-beds to treat contaminated waters off-site.

Composting is a variant of biological treatment which is also suitable for contaminated soils but has not, as yet, been widely used. Here, large quantities of organic matter such as sewage sludge or green manure are mixed with the contaminated soil, moisture levels are high and the high level of microbial activity which results causes a rise in temperature. If contaminants are reduced to an acceptable level the resultant material is likely to provide an excellent growing medium to be returned to site.

Biological treatment for contaminated soils has been one of the more widely used of the decontamination techniques, and North American, German and Dutch systems, particularly for off-site treatment, have become quite sophisticated.

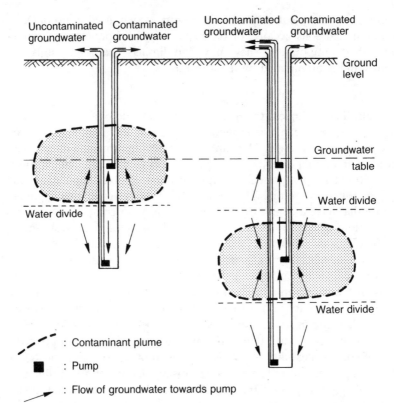

Uncontaminated groundwater   Contaminated groundwater   Uncontaminated groundwater   Contaminated groundwater

Ground level

Groundwater table

Water divide

Water divide

Water divide

Water divide

: Contaminant plume

: Pump

: Flow of groundwater towards pump

Fig. 4.10   Dual pumping system for groundwater treatment (source, Richards *et al.*, 1993).

4.4.4   *Ground improvement techniques*

In addition to overcoming the constraints of contamination, reclamation techniques also have to provide suitable ground conditions for new uses of the site. For 'hard' end-uses the techniques for providing suitable conditions for construction are known as ground improvement techniques. The principal aim of these techniques is to decrease the variability in the physical conditions of the ground and provide enough strength to support new construction on-site. Table 4.10 provides details of some of the ground improvement techniques in use.

Many industrial sites are in areas of former mining and the ground may be unstable because of underground workings, or may comprise combustible coal or colliery spoil. The treatment of such land is beyond the scope of this book but the methods employed may be found in Richards *et al.* (1993).

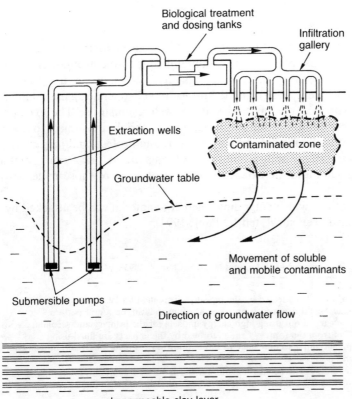

Fig. 4.11 Example of an idealized recirculating *in situ* biological treatment programme (source, Richards *et al.*, 1993).

4.4.5 Landfill sites

Industrial sites may be found on or near former or active landfill sites. The characteristics of landfill sites are summarized in Table 4.11. Principal concerns for redevelopment will be migration of landfill gas and leachate, and settlement of fill materials. The site assessments described in section 4.3 should characterize the site with respect to all of these parameters; in addition to taking measures to deal with the contaminants on-site the proximity of landfill site may mean that long-term measures need to be taken to protect site and buildings from the effects of gas and leachate. These measures will include barriers and trenches described in section 4.4.2 but may also require measures being taken at the construction stage such as installation of passive or active gas venting systems in buildings and the operation of gas alarms.

4.4.6 *Protection of construction materials*

At reclaimed sites consideration will need to be given to the protection of construction materials from attack by contaminants in the ground. Guidance has been issued on the protection of concrete for many years (Building Research Establishment, 1991), and the protection of concrete against sulphate attack, which can occur in natural soils, is well understood. The extent to which and the conditions under which building materials are subject to attack by substances other than sulphate are less clear, and professional judgement is important in assessing whether building materials are at risk from contaminants in the ground and the extent to which protection of those materials is needed.

Generic conditions which determine whether a material will be attacked by a contaminant have been suggested by Paul (1994):

Table 4.10 Ground improvement techniques (source: Charles, 1993)

Dynamic compaction

Deep compaction of the fill is effected by the repeated dropping of a heavy weight using a crane, e.g. a 15 tonne weight dropped from a height of up to 20 m, on a grid pattern.

Treatment may be carried out using high-energy impacts for the primary and secondary grids, the latter being offset from the former. Craters formed by the impacts are filled using earthmoving equipment and a second stage of more uniform treatment is carried out using a reduced drop height.

An alternative method of treatment for shallower fills is to use a rapid impact compactor, which involves dropping a 7 tonne weight from a height of 1.2 m onto a circular plate.

Vibro techniques

Vibro techniques involve the compaction of granular soils or the formation of stone columns using a vibrating cylindrical poker suspended from a crane. The poker may be 300–450 mm diameter and can weigh up to 4 tonnes. Treatment depths of around 6 m are typical, but depths up to 30 m have been achieved. The poker penetrates the ground as a result of the vibratory action assisted by flushing jets in the nose cone and sides. Compressed air is generally used for flushing, although water may sometimes be employed. Vibro treatment is carried out on a grid pattern, using closer grid spacings where higher bearing capacities are required, e.g. beneath pad foundations or edge beams.

'Vibrocompaction' is the term generally used to describe the densification of granular soils using vibro techniques. Additional material may be introduced from the top of the hole as part of the compaction process. An alternative in unstable ground is to use the bottom feed process whereby stone is introduced directly from the top of the poker via a feed pipe.

'Vibrated stone columns' is the term used for the stone columns introduced and compacted by the vibro process in order to improve the bearing capacity.

Preloading

Preloading involves over-consolidation of the fill by temporary surcharging to improve the bearing capacity prior to construction taking place. Surcharging is usually carried out using several metres of fill material placed using earthmoving equipment and left *in situ* over a period of two to three months.

Excavation and recompaction

This method involves the treatment of loose granular fill material by excavation and recompaction in thin layers under controlled conditions using conventional earthmoving equipment.

1. The presence of water, except for free-phase organic compounds.
2. The availability of the contaminant in terms of concentration and replenishment rate.
3. Contact between the contaminant and the building material.
4. The sensitivity of the material to the contaminant.

Draft exposure classes for attack of concrete by substances other than sulphate have also been proposed (European Committee for Standardization, 1990). Some effects of contaminants in the ground on building materials are summarized in Table 4.12. In addition to the chemical agents listed, microbial activity can also affect building materials. The most common microbial agents are the sulphur-reducing bacteria *Thiobacillus* and *Desulphovibrio*, both of which catalyse the formation of sulphuric acid which affects both cement-based and metal products. Plastics can be attacked by bacteria and fungi, although polyethylene and PVC are generally resistant.

Table 4.11   Landfill sites

Characteristic	Implications for reclamation
Waste type	
Inert waste fill	Contains demolition rubble and excavated material unaltered from their natural state. Can contain biodegradable material (e.g. wood, paper). Could produce leachate and landfill gas, particularly if control over materials filled was not good.
Household waste	Domestic refuse. Contains large amounts of biodegradable material which, depending on materials and conditions within the landfill, could biodegrade over many years to produce landfill gas and leachate.
Commercial waste	Arising from commercial premises. Likely to contain large amounts of paper. Can produce landfill gases and leachate.
Industrial waste	Waste from industrial processes. Can be putrescible and depending on the nature of the waste produce toxic leachates.
Special waste	The most toxic of wastes usually from industrial processes. Can be putrescible and produce toxic leachates.
Age of tip	
Old	The older the tip, the less likely there will be good records of tipped materials. Tips may produce landfill gas for over 30 years and gas production should always be tested for. Leachate may be produced for much longer periods than landfill gas, although leachate characteristics will change as biodegradation proceeds. Very old domestic tips (pre-1940) contained much more inert waste than modern tips – less putrescible material was thrown away and tips often contain a high proportion of ashes.
Young	Recently constructed tips are well designed and the waste they receive is strictly controlled. Leachate production from modern tips is rare and landfill gas production is controlled and often used as a source of energy. Tip restoration is carried out as part of tip management.

Table 4.12   Effects of contaminants on building materials (after Paul, 1994)

Material	Contaminant	Mechanism and effect
Concrete	Sulphate	1. **Formation of gypsum** Calcium hydroxide ($Ca(OH)_2$) is replaced by Gypsum ($CaSO_4 \cdot 2H_2O$) resulting in expansion, leading to deformation and cracking. 2. **Formation of ettringite** In the presence of aluminates, calcium hydroxide reacts with sulphate to form the sulphoaluminate ettringite ($CaO \cdot Al_2O_3 \cdot CaSO_4 \cdot 32H_2O$), resulting in expansion. Although the precise reason for the expansion is unclear it is likely to be due in part to the large amount of water of crystallization in ettringite. 3. **Formation of thaumasite** Thaumasite ($CaCO_3 \cdot CaSO_4 \cdot CaSiO_3 \cdot 15H_2O$) forms a solid series with ettringite, contains no aluminium but does contain silicates. It can be found without ettringite. It is only formed quickly at low temperatures and its formation results in expansion.
	Chlorides	1. **Degradation of cement binder** Chloride reacts with calcium hydroxide to form calcium chloride which leaches away, causing the concrete to become more permeable, allowing ingress of other chemicals. 2. **Salt crystallization** In the presence of chloride, wetting and drying cycles due to groundwater fluctuations can cause chloride salt crystal formation within the concrete pores. If crystallization pressure is greater than the tensile strength of the concrete, then cracking and disintegration will occur. 3. **Formation of expansive reaction products** In a similar way to ettringite formation chloroaluminates (e.g. $CaO \cdot Al_2O_3 \cdot CaCl_2 \cdot 10H_2O$) can be formed resulting in expansion.
	Acids	Acids attack cement hydrates, particularly calcium hydroxide, resulting in the formation of soluble salts which leach out allowing ingress of other chemicals. Strong acids such as sulphuric and hydrochloric can attack the cement binder, resulting in disintegration of the concrete and can also produce expansive reactions.
	Magnesium salts	Magnesium salts (except magnesium hydrogen carbonate) cause the calcium in the cement binder to be replaced by magnesium, forming soluble calcium salts which are leached out. The binder loses power and the concrete disintegrates.
	Ammonium salts	Ammonium acts in a similar way to magnesium by replacing calcium in the cement binder. Ammonium antagonizes the corrosive action of associated anions such as sulphate by increasing the porosity of the cement paste thereby allowing more rapid attack by sulphate.
	Organic compounds	Most organic compounds do not attack hardened concrete but can affect the hardening of fresh concrete. **Hydrocarbons** Hydrocarbons entirely of mineral origin do not attack concrete but vegetable oils can oxidize to form acids which cause concrete degradation. Hydrocarbons readily seep through concrete however and if the hydrocarbons contain aggressive substances then the hydrocarbon ingress will cause more rapid concrete attack. **Phenols** Phenol is a weak acid which will react with calcium hydroxide to form calcium phenolate. Crystallization of calcium phenolate is an expansive reaction causing deterioration of the concrete.

Table 4.12 (*contd.*)

Material	Contaminant	Mechanism and effect
		Alcohols Alcohols absorb water from concrete, reducing its strength. Some alcohols such as glycol act as a weak acid and can react with calcium hydroxide forming soluble calcium salts which are leached out.
Reinforced concrete	Chloride	Corrosion of steel reinforcement in concrete is stimulated by the presence of chloride ions even in highly alkaline conditions where the reinforcement is usually safe from the electrochemical reactions which cause steel reinforcement corrosion. Corrosion of reinforced concrete in the ground is most likely at saturated/unsaturated and soil/atmosphere interfaces because these conditions are likely to increase the porosity of concrete to chloride ions.
Asbestos cement	Sulphate	Although asbestos cement is less permeable than concrete this does not in practice increase its resistance to sulphate attack. In an aggressive sulphate environment asbestos cement has the same susceptibility as any other cementitious product.
Mortar	Sulphate	Mortar is as susceptible to sulphate attack as any other cementitious product.
Bricks	Acids and alkalis	Bricks are generally resistant to deterioration in acid conditions with the exception of hydrofluoric acid. In basic conditions the silica in the bricks forms soluble hydroxides which can be leached away.
	Salts	Soluble salts (e.g. sodium or calcium sulphate) can migrate into brickwork and then crystallize at the surface through evaporation. Although not damaging, salt crystallization can be unsightly. Porous bricks are the most affected. Calcium silicate bricks are resistant to attack by most sulphates but are susceptible to high concentrations of ammonium and magnesium sulphate and can be affected by a combination of strong solutions of calcium or sodium chloride and frost.
Cast iron	Acids	Under acidic conditions soluble salts of iron will form and be leached away.
	Salts	The effects of salts depend on the ions present and whether conditions are oxidizing and acid. In these conditions corrosion may occur particularly if sulphate and chloride are present. At neutral pHs, however, the presence of salts can cause a protective layer to form on the surface of the iron.
Steel		Steel corrodes in aerobic conditions and negligible steel corrosion has been found in undisturbed soils. In the heterogeneous conditions of many contaminated sites and, particularly if aerated fill materials are present, corrosion may occur.
Copper		Copper is generally resistant to corrosion because of the protective film of copper oxide which forms. In oxidizing conditions it can be attacked by acids (e.g. nitric acid) and strong alkalis. In moist conditions sulphate and chloride can cause corrosion. Cyanides may also attack copper.
Polyethylene	Acids	Oxidizing acids such as sulphuric and nitric acid are highly corrosive to polyethylene. Hydrochloric acid is not corrosive but can weaken polyethylene by diffusing into it and weakening intermolecular chains.

Table 4.12 *(contd.)*

Material	Contaminant	Mechanism and effect
	Alkalis	Alkaline substances such as basic salts, ammonia solutions and bleaching chemicals cause deterioration of polyethylene through stress cracking.
	Organic compounds	Organic compounds do not attack polyethylene although some can cause swelling or stress cracking. Organic compounds of low molecular weight (e.g. benzene and toluene) will, however, permeate polyethylene.
Polyvinyl chloride (PVC)	Acids	Oxidizing acids degrade PVC, particularly nitric acid which causes it to become brittle.
	Hydrocarbons	PVC has generally good resistance to hydrocarbons. Chlorinated hydrocarbons, anilines, ketones and nitrobenzene cause PVC to swell and eventually dissolve. These compounds also permeate PVC.
Rubbers	Acids and salts	Rubbers are resistant to non-oxidizing acids but are attacked both by oxidizing acids (e.g. nitric) and oxidizing salts of metals such as copper, manganese and iron.
	Hydrocarbons	Hydrocarbons attack rubber causing it to swell and soften although different rubbers have different levels of resistance to specific hydrocarbons. Hydrocarbons will also permeate rubber.

The resistance of rubbers to microbial attack varies with natural rubbers being more susceptible than synthetic rubbers.

4.5 Conclusions

There have been major advances in the treatment of land subject to contaminative use, as can be seen from the above. There still remains room for development of these techniques, such as in the treatment of heavy metal wastes not involving major destruction of the structural integrity of the soil substrate.

At the end of the processes described up to now, the next phase of reclamation or restoration can proceed, that of revegetation of the 'clean' substrate. That is the focus of the next two chapters.

References

BRITISH STANDARDS INSTITUTION (1988). *Code of Practice for the identification of potentially contaminated land and its investigation*. Draft for development DD175. British Standards Institution, London.

BUILDING RESEARCH ESTABLISHMENT (1991). *Sulphate and acid resistance of concrete in the ground*. BRE Digest 363. Building Research Establishment, Garston.

CANADIAN COUNCIL OF MINISTERS OF THE ENVIRONMENT (1993). *Guidance manual on sampling, analysis and data management for contaminated sites*. CCME, Winnipeg, Manitoba, CCME-EPC, NCS62E.

CHARLES, J.A. (1993). *Building on fill: geotechnical aspects*. Building Research Establishment, Garston.

DEPARTMENT OF THE ENVIRONMENT (1994a). *Guidance on the preliminary site inspection of contaminated land*. Contaminated Land Research Report No. 2. Department of the Environment, London.

DEPARTMENT OF THE ENVIRONMENT (1994b). *Sampling strategies for contaminated land.* Contaminated Land Research Report No. 4. Department of the Environment, London.

DEPARTMENT OF THE ENVIRONMENT (1994c). *The reclamation and management of metalliferous mining sites*. HMSO, London.

EUROPEAN COMMITTEE FOR STANDARDIZATION (1990). *Concrete – performance, production, placing and compliance criteria*. European pre-standard env 206. CEN, Brussels.

FERGUSON, C.C. (1992). The statistical basis for spatial sampling of contaminated land. *Ground Eng.*, **June 1992**, 34–38.

ICRCL (1987). *Guidance on the assessment and redevelopment of contaminated land*. ICRCL59/83 2nd edition. Department of the Environment, London.

NATIONAL RIVERS AUTHORITY (1994). *Leaching tests for the assessment of contaminated land: interim NRA guidance*. R&D Note 301. National Rivers Authority, Bristol.

NEDERLANDS NORMALISATIE-INSTITUUT (1991). *Soil: investigation strategy for exploratory survey*. Draft standard UDC 628.516. NVN 5740. Nederlands Normalisatie-Instituut, Delft.

PAUL, V. (1994). *The performance of building materials in contaminated land*. Building Research Establishment, Garston.

RICHARDS, I.G., PALMER, J.P. and BARRATT, P.A. (1993). *The reclamation of former coal mines and steelworks*. Elsevier, Amsterdam.

UNITED STATES ENVIRONMENTAL PROTECTION AGENCY (1992). *Characterising heterogeneous wastes: methods and recommendations*. EPA/600/R-92/033. US Environmental Protection Agency, Las Vegas, Nevada.

WELSH DEVELOPMENT AGENCY (1993). *The WDA manual on the remediation of contaminated land*. WDA, Cardiff.

WELSH OFFICE (1988). *Survey of contaminated land in Wales*. Welsh Office, Cardiff.

Further reading

BULLOCK, P. and GREGORY, P.J. (1991). *Soils in the urban environment*. Blackwell Scientific, Oxford.

CAIRNEY, T. (ed.) (1987). *Reclaiming contaminated land*. Blackie, Glasgow.

LEVIN, M.A. and GEALT, M.A. (ed.) (1993). *Biotreatment of industrial and hazardous waste*. McGraw-Hill, New York.

SMITH, M.A. (ed.) (1985). *Contaminated land: reclamation and treatment*. Plenum, New York.

Section 3

Restoration of reclaimed land

Chapter 5

Preparing the substrate for growth and reclamation to agro-forestry uses

5.1 Introduction

During the course of this chapter we will consider situations where there is no *adverse chemical impediment* to plant growth, i.e. the substrate is uncontaminated by toxic compounds. This does not exclude those sites where the substrate is poor in nutrients and structure, owing to its origin or treatment during the industrial activity, and also includes sites where no direct use has been made of the substrate, but where it has been subject to handling or storage.

We must also recognize that in many circumstances the original substrate has been lost. The topsoil may have been sold off at the start of the works, or was such a thin cover as the system was a pioneer or naturally stressed system. Many older works, such as clay-mines in south-western England, have no records of where the soil has gone. In the light of the changes occurring in the soil during storage (Chapter 3), it is unlikely that soils stored for 100 years or more would be of much use in any case. Preparing for growth in these types of substrate will be covered later in this chapter. We shall also be considering reclamation to agriculture and forestry, as although these may be end-points in themselves, they are often the starting point for restoration schemes. Restoration to such self-sustaining and amenity uses will be covered in Chapter 6.

When facing the task of devising a restoration management plan for a new site, two main approaches may be taken: the prescribed and the experimental.

The *prescribed approach* is where a blanket set of recommendations and *a priori* management techniques are employed. This is based on the experience of working on similar, and sometimes dissimilar, reclamation and restoration schemes. The prescribed approach is extremely inflexible when new problems arise, particularly when the substrate or hydrogeology is not what was expected, owing to the fact that all of the planning (both technical and financial) has been made before the site was disturbed.

The *experimental approach* allows the site and materials to be investigated in a much more flexible manner, and is therefore more expensive. The experimental approach has itself further subdivisions, those designed to find out whether something will work, e.g. a particular fertilizer regime, and those designed to find out how something works. This approach was pioneered by Anthony Bradshaw at the University of Liverpool, and puts restoration at the very centre of ecological thinking and research, as the systems involved are being tested to their limits, and large forces are being laid bare.

The majority of restorations carried out are in the former category, and may explain why many of them fail. It must be remembered, however, that the managers are constrained by an inflexible planning structure, and often achieve success in the face of such adversity.

5.2   Substrate handling and reinstatement procedures

At this point conditions of anthropogenic toxicity may not exist, but there will still be some obstacles to be overcome:

1. Lack of a suitable substrate.
2. Lack of nutrient supply, through absence or being present in a plant-unavailable form.
3. Poor substrate structure or structural instability.
4. Inadequate or excess supply of water.
5. Incorrect landform.
6. Lack of biological potential, in one, two or all of the functional groups (cf. Chapter 1).

At this point the question of substrate suitability must be addressed and the flowchart shown in Fig. 5.1 may be used to make decisions over the first step, substrate handling and treatment.

Much of the same equipment used in the construction of top- and subsoil stores may be used in reinstatement of soil profiles after the original contours have been returned. The angle of the final slope will have a significant effect on the type of land-use which may be carried out upon it subsequently, with forestry suitable up to 35°, pasture 15° and crops 5°.

This may involve the movement of large volumes of overburden material, i.e. bedrock not containing the mineral resource. The overburden is usually moved on a continuous basis whilst the site is being worked, by combinations of draglines and dumper trucks. The surface of the overburden is then ripped by a bulldozer pulling large tines through the material to a depth of 8–20 cm, to prevent the formation of an iron pan, by encouraging the forma-tion of drainage channels between the overburden and subsoil. When the over-burden has been replaced there is a site meeting between the interested parties to agree that these are satisfactory; this includes a representative of the site contractor, a site engineer for British Coal in the case of open-cast coal-mining, the site owner, local authority officer, Ministry of Agriculture, Fisheries and Food (MAFF) officers, and members of local interest groups.

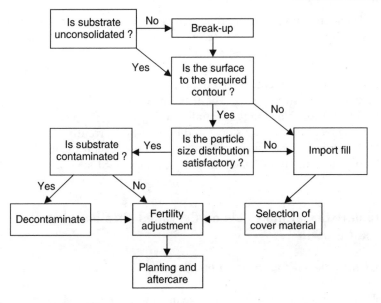

Fig. 5.1 Substrate decision flowchart.

5.2.1 Soil reinstatement

Subsoil is removed from stores, usually by means of an earthscraper. Subsoil is reinstated in two layers, giving a final minimum thickness of 60 cm, where an agricultural end-use is intended. This process is followed by the reinstatement of the topsoil, to a minimum depth of 25 cm. This again is usually carried out on large-scale sites by means of earthscrapers, but use of dump truck and backhoe is becoming the preferred method to minimize damage, with the topsoil being replaced in strips 3–5 m long, approximately 1.5–2 m wide. In many cases where topsoil and subsoil resources are not available alternative substrates or 'soil-forming materials' may be used.

A process of 'subsoiling' is then carried out. A tractor pulls a plough with multiple winged tines through the layer where the topsoil meets the subsoil. Ripping is done to encourage drainage between the two layers. In the case of open-cast coal-mining, when all the soils in an agreed phase have been replaced a site meeting is held, when, if agreement is reached, the land is released into the next phase of management.

At the end of this process the soil or substrate will be compacted with no structure at almost any level, and as a result will be prone to waterlogging and extreme droughtiness, as its water-holding capacity will have been decreased and its bulk density increased (Fig. 3.3). We must first consider the situation where there is no original soil available, topsoil may be imported from off-site, or a replacement soil-forming material found.

There has been an innovative approach to limestone quarry restoration taken by the Limestone Quarry Reclamation Group at Manchester Metropolitan University. On inspecting old abandoned quarries it became clear that many were rich in biological, substrate and landform diversity, through the action of the elements on old quarry faces leading to the formation of slip slopes between buttress areas. The team at Manchester mimicked this landform by 'restoration blasting'; essentially taking newly finished quarries and blasting along faces to produce slopes and buttresses. Currently research is under way to determine the best means of introducing the vegetation and fauna found on the old sites, many of which have now gained SSSI status.

5.3   Characteristics of materials on degraded sites as substrates for plant growth

5.3.1   Types of material found on industrial sites

The principles of soil biology and nutrient cycling and their methods of determination were described in Chapters 1 and 2. Here the extent to which these principles can be applied to materials found on degraded sites will be discussed, along with a consideration of them as potential substitute substrates on sites where the original material has been lost. The range of materials to be found on industrial sites is very great and can vary considerably within one site. In Fig. 5.2 characteristics of some of the typical wastes found at industrial sites are listed with respect to the constraints they present to plant growth. Nutrient availability is a factor on all substrate types except domestic refuse and even here it can sometimes be a problem, depending on the precise nature of the waste. Toxicity is a problem with half the site materials.

A third of the derelict land in the United Kingdom comprises spoil heaps and about 20 per cent is industrial dereliction (cf. Chapter 1). Of the spoil heaps, colliery spoil heaps and metalliferous mine spoil make up 12 per cent each with china clay heaps making up a significant proportion of the rest. Consideration of the characteristics of these three spoil materials will provide an indication of the constraints on vegetation establishment on the majority of British spoil heaps. Derelict industrial sites are less easy to characterize because of the range of industrial activities which have been carried out. However, most derelict non-mining industrial sites have a predominance of two types of material:

1. Demolition rubble comprising predominantly bricks, mortar and concrete.
2. Fill material which may comprise demolition rubble but often includes wastes from industry and waste material from engineering operations. In coal-mining areas a significant proportion of filled land will be filled with colliery spoil.

	Stability	Combustion	Slope angle	Flooding, stress	Toxicity	Compaction	Temperature	Wind erosion	Nutrients	Stoniness	Uneven surface	Erosion	Soil fauna and microbes
Colliery spoil	✔	✔	✔	✔	✔	✔	✔	✔	✔	✔	✔	✔	✔
Smelter slag	✔	✔	✔	✔	✔	–	✔	✔	✔	✔	✔	✔	✔
Slate/shale	✔	–	✔	✔	–	–	✔	✔	✔	✔	✔	✔	✔
Metal wastes	✔	–	✔	–	✔	✔	✔	✔	✔	✔	✔	✔	✔
Quarry pits	–	–	✔	✔	–	✔	✔	✔	✔	✔	✔	✔	✔
Brick pits	–	–	✔	✔	–	✔	✔	✔	✔	✔	✔	✔	✔
China clay	✔	–	✔	✔	–	–	✔	✔	✔	–	–	✔	✔
Ironstone	✔	–	✔	✔	–	✔	✔	✔	✔	–	✔	✔	✔
Chemical waste	✔	–	–	–	✔	✔	✔	✔	✔	–	–	✔	✔
Pulverized fuel ash	–	–	–	✔	✔	✔	✔	✔	✔	–	–	✔	✔
Sand and gravel	–	–	✔	✔	–	✔	✔	✔	✔	–	✔	✔	✔
Domestic refuse	–	✔	–	✔	✔	✔	–	✔	–	–	✔	–	–

Fig. 5.2 Constraints on vegetation development on different wastes (after Kent, 1982).

Many former industrial sites may also have soil material, original or imported, of a range of qualities. Original soil may have been damaged by compaction, waterlogging or pollutants. There may be poor subsoil or a mixture of subsoil or topsoil and waste materials or soils developed from waste materials over a long period of time. A tentative classification of soils occurring in urban areas is given in Table 5.1.

The restoration of quarries after mineral extraction is a significant activity and this will be considered with particular reference to limestone quarries. Table 5.2 gives the characteristics of colliery spoil, metalliferous mine spoil, china clay waste, demolition rubble and limestone waste relevant to plant growth. Those characteristics that have major implications for revegetation are: extremes of pH; toxicity; lack of plant nutrients; low organic matter content; particle size distribution; and compaction or consolidation.

pH Substrate pH influences plant growth mainly through its effect on the solubility of chemical elements, including those that are directly toxic to plants and those that are required as nutrients. Figure 5.3 shows the relationship

Table 5.1   A classification of urban soils (after Gilbert, 1989)

1. Man-made humus soil. A thick (> 40 cm) man-made A horizon resulting from bulky amendments of manure, mineral matter or domestic rubbish. Allotments, old gardens, small holdings, rubbish dumps.
2. Topsoiled sites. A/C or A/BC soils created by spreading topsoil of variable quality over raw, disturbed, mineral soil. May be compacted at one or more levels. The evolution of these rankers to brown soils is helped by cultivation. Recently landscaped sites.
3. Raw-lithomorphic soils. Initially these consist of little altered raw mineral materials, often of man-made origin, at least 30 cm deep, and show no horizon development attributable to pedogenic processes. With time they evolve into lithomorphic soils as a distinct humose upper layer develops. Two subdivisions can be made in each group: (1) soils with only one layer that is not humified topsoil; (2) soils with at least two layers including a humified topsoil.

   (a) Brick rubble. Man-made raw soil evolving into pararendzina. Brick rubble often mixed with a little old soil and subsoil. Building demolition areas.
   (b) Furnace ash, slag, cinder. Man-made raw soil evolving into humic ranker. A coarsely textured non-calcareous soil associated with railway land, particularly old sidings, heavy industry, etc.
   (c) Chemical wastes. Man-made raw soil evolving into a range of lithomorphic soils. The raw material may have an extreme pH or contain highly toxic substances. Of limited occurrence, associated with the chemical industry, certain other manufacturing industries and electricity generation (pulverized fly ash).
   (d) Disturbed subsoil. Man-made raw soils evolving into profiles that have affinities with ranker-like alluvial soils in which the humose layer passes down into an unconsolidated C horizon which may show stratification. These soils form from a stratum of artificially rearranged or transported material > 40 cm thick consisting predominantly of mingled subsoil horizons (B, B?C or C). Found in some recently landscaped sites, some landfill sites.
   (e) *In situ* subsoil. Raw skeletal soil evolving into a variety of lithomorphic soils. Unconsolidated or weakly consolidated mineral horizons, sometimes partly stratified, which have weathered *in situ*. Found on steep banks associated with transport corridors or sites benched for major buildings. Also widespread in non-urban areas.
   (f) Hard surfaces. Tarmacadamed, paved and concrete surfaces. Roads, pavements, floors of demolished factories, angle between pavement and wall, etc.

between pH and the availability of the major plant nutrients. At pH 6.5 nutrient availability to plants is at a maximum and toxicity at a minimum. Soils of lower pH are commonly found supporting commercial forestry, extensive grazing or semi-natural vegetation, and a limited range of plant species will colonize very acidic substrates of pH 3.5 or lower. However, at low pH plant establishment is limited by the concentrations of toxic element in solution. So, for example, peat soils at pH 3.5 are often well vegetated whereas mineral substrates such as colliery spoil are not. Few leguminous plants are well adapted to acidic soils and nutrient availability may be limited by immobilization of nutrients in such soils. The production of self-sustained swards therefore requires the correction of excessive acidity, and/or the selection of vegetation tolerant of the acid conditions which exist.

Toxicity   Some derelict land materials can contain substances highly toxic to plants. For example, metalliferous spoils can contain many tens of thousands

Table 5.2 Characteristics of waste materials with potential use as plant growth substrates

Item	Colliery spoil	China clay waste	Metalliferous mine waste	Demolition rubble	Limestone waste
pH	<3 to >8	<4 to 6	<3 to 8	7–10	7.5–8.5
Physical properties	Weathering silt and mudstones	Mica (silt) or sand waste, the latter free-draining	Clay to gravel size, thixotropic to free-draining	Can be blocky but often good size gradation. Free-draining	Can be blocky with a proportion of fines under the surface. Tendency to cement if large amount of fines
N	Nitrogen-deficient	Nitrogen-deficient	Nitrogen-deficient	Nitrogen-deficient	Nitrogen-deficient
P	Phosphate-deficient, fixes phosphate	Phosphate-deficient, easily leached out	Phosphate-deficient	Some phosphate	Phosphate-deficient
K	Adequate supply	Deficient, easily leached out	Deficient	Some potassium	Deficient
Organic matter	Deficient	Deficient	Deficient	Deficient	Deficient
Potential toxicities	Al and Zn at low pH, salinity on some sites	None	Heavy metals	None unless contaminated by other substances	None
Moisture	Waterlogged in winter, potentially droughty in summer	Sand waste: low moisture-holding capacity. Mica waste: less of a problem	Sands and gravels have low moisture-holding capacity; silts retain moisture	Variable but porous brick can retain plant available moisture	Low moisture level a problem in absence of fines
Erodibility	Easily eroded	Easily eroded	Easily eroded	Not usually a problem	Not usually a problem
Compaction	Can be compacted	Not compacted	Not compacted	Can be compacted	Can be compacted and cementitious

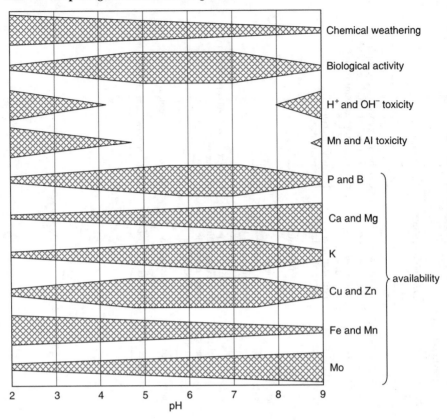

Fig. 5.3 The influence of pH on nutrient availability (after Bradshaw and Chadwick, 1980).

of parts per million of heavy metals which restrict the growth of all but the most tolerant of plants (see section 5.5.2), chemical waste may contain chemicals which exclude the growth of all plants, the decomposition of domestic waste may result in the production of organic compounds toxic to plants and their dispersal in leachate. In some wastes the nature of toxicity to plants may change over time. For example, pulverized fuel ash (PFA), a waste product of electricity generation from coal, can contain initially high concentrations of boron which restrict plant growth but which then leach out, reducing the toxicity of the substrate. Similarly the wastes from nineteenth century alkali production, such as Leblanc waste, deposited in large heaps in north west England, had pHs on deposition as high as 12–14 and high concentrations of sulphate, restricting many plants. Over time, sulphate and pH levels dropped, allowing colonization often by ecologically valuable plants. If these materials are excavated or reshaped in some way, previously buried unweathered high pH material may be brought to the surface. At some sites a mosaic of toxicities will occur because many different types of materials are found together on one site.

Toxicity is not, however, a simple matter of particular concentrations of substances being toxic to a plant. The chemical form in which potentially toxic elements are found, the presence of other chemicals which may aggravate or ameliorate the toxicity of a particular chemical, the prevailing pH and nutrient status will affect the way plants respond to such substances in the soil.

Lack of nutrients   Nitrogen and phosphorus are found at extremely low plant available concentrations in most degraded land materials. Both nutrients are essential for the growth of plants and when supplies are inadequate young plants will fail to establish whilst established plants will become moribund and decline (Fig. 5.4). Legumes are particularly sensitive to a lack of available phosphate and the fixation of atmospheric nitrogen in the absence of adequate phosphate supplies will be poor. The effect of pH and microbial substrate availability on the transformation of nitrogen into a plant-available form was demonstrated on colliery spoil by Williams (1975) (Fig. 5.5). The principles underlying these effects are relevant to all degraded land materials where nitrogen is in short supply:

1. If pH is low, nitrogen remains in the ammonium form and is not microbially converted to the more plant-available nitrate.
2. If the pH is raised, ammonium is readily converted to nitrate and available to plants.

Low pH also prevents other microbial activities such as decomposition of plant

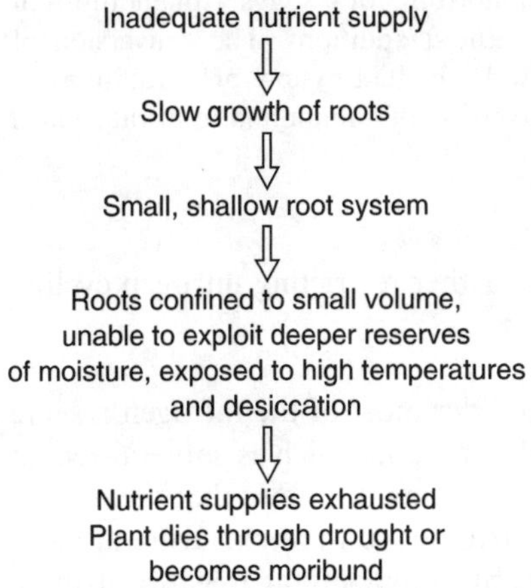

Inadequate nutrient supply
⇩
Slow growth of roots
⇩
Small, shallow root system
⇩
Roots confined to small volume,
unable to exploit deeper reserves
of moisture, exposed to high temperatures
and desiccation
⇩
Nutrient supplies exhausted
Plant dies through drought or
becomes moribund

Fig. 5.4   The consequences of inadequate nutrient supply for plant growth (source, Richards *et al.*, 1993).

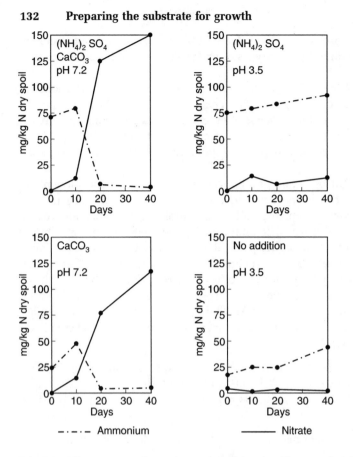

Fig. 5.5 Nitrogen transformations on reclaimed colliery spoil (after Williams, 1975). The colliery spoil was incubated at 25°C and 10% moisture for 40 days with additions of ammonium sulphate or calcium carbonate or without additions. The conversion of ammonium to nitrate is microbially controlled and inhibited by low pH. Additions of calcium carbonate raise the pH and facilitate conversion of ammonium to nitrate which is then available to plants.

material and mineralization to ammonium, further restricting nitrogen cycling (Table 5.3, Fig. 5.6).

Lack of organic matter   Organic matter provides most of the nitrogen reserve in soils and comprises typically 5 per cent nitrogen which is mineralized at about 2 per cent per year. If organic matter is lacking it follows that the reserve of nitrogen is also poor. Organic matter contributes to the structuring of soils, particularly those with a high clay content, by stabilizing aggregates of these fine particles. Poorly structured soils, such as colliery spoil, will consolidate as they weather and the clay content increases. Building rubble, however, will weather very slowly and remain coarse-grained with large pore spaces.

Table 5.3   Nitrogen cycling

In order for nitrogen to be cycled it has to be in compounds where the carbon to nitrogen ratio is low enough for microbes to be able to break them down. The critical carbon to nitrogen ratio is between 25 and 30 to 1 and above this level, i.e. where there is a greater proportion of carbon, microbial breakdown of organic matter may not occur.

Many grass, herb and tree roots and the woody and dead parts of plants have a carbon to nitrogen ratio higher than 30:1 and in order to break these materials down microbes need a supply of nitrogen from elsewhere. In fertile soils this nitrogen is available so materials of a high carbon to nitrogen ratio are broken down and the nitrogen becomes available to be taken up by plants again. In infertile soils there is no freely available nitrogen and nitrogen can become bound up in compounds of high carbon to nitrogen ratio or microbial biomass.

Other mechanisms may cause resistance to microbial breakdown of organic matter, e.g. (after Stevenson, 1982):

1. Stabilization of proteinaceous constituents (e.g. amino acids, peptides, proteins) through their reaction with other organic soil constituents (e.g. lignins, tannins, quinones, reducing sugars).
2. The formation of biologically resistant complexes by the chemical reaction of NH_3 or NO_2 with lignins or humic substances.
3. Protection of organic N compounds from decomposition by their absorption on to clay minerals.
4. Stabilization of organic N compounds by the formation of complexes with polyvalent cations.
5. The siting of organic N in pores or voids physically inaccessible to micro-organisms.

The significance of these mechanisms will be greater in infertile soils.

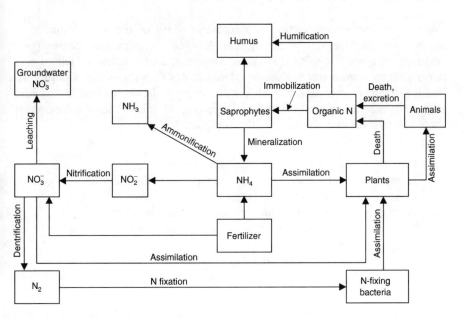

Fig. 5.6   A nitrogen cycle applicable to derelict land substrates.

Consolidation of substrates can lead to impeded drainage, poor available water capacity and the restricted root extension of plants where organic matter content is low. Colliery spoil is particularly prone to such consolidation effects. As a result, extremes of summer drought and winter waterlogging are common on colliery spoils. Organic matter also absorbs moisture and so directly improves the available water capacity of the spoil.

Particle size distribution   Substrate particles larger than 2 mm retain very little water and where these are abundant they can significantly dilute the finer-grained soil. The available water capacity of soil is related to its texture. Very coarse-grained substrates have few soil pores of the size needed to hold water against gravity drainage. Coarse-grained materials will therefore be excessively free draining and plants with roots confined to the upper layers will be subject to extreme water stress. However, once deep-rooting plants, particularly trees, become established, and their roots extend below the zone of greatest soil-drying, even very coarse-grained materials can support a vegetation cover.

Very fine-grained materials on the other hand can become saturated and anaerobic and provide difficult conditions for plant establishment. Surfaces may be dry, dusty and exposed giving seedlings little shelter and the underlying saturated zone with anaerobic deeper layers provide difficult rooting conditions and little root anchoring potential.

Compaction   Spoil compaction is a common result of the use of mining or engineering machinery to place or regrade spoils. Compaction causes poor drainage and waterlogging during wet weather, and creates a lack of soil pores able to store water for use by plants in dry conditions (see Chapter 3). Plants growing in compacted spoils produce very shallow root systems because of the physical impenetrability of compacted spoil and the seasonal death of roots in the saturated zone above the compacted layer.

As a result, the root system:

1. Provides poor physical anchorage for the plant.
2. Is confined to the zone of greatest temperature fluctuation.
3. Is confined to the zone of greatest moisture stress.
4. Is unable to exploit nutrients beyond the surface layer.
5. Is unable to exploit nutrients in the surface layers during drought or waterlogged periods.

5.3.2   Treatment considerations

In order to establish vegetation on these materials the various constraints described must be overcome. One way to overcome constraints would be to use a covering system (see Chapter 4) to provide a soil material in which vegetation can establish. However, covering systems are costly and unnecessary

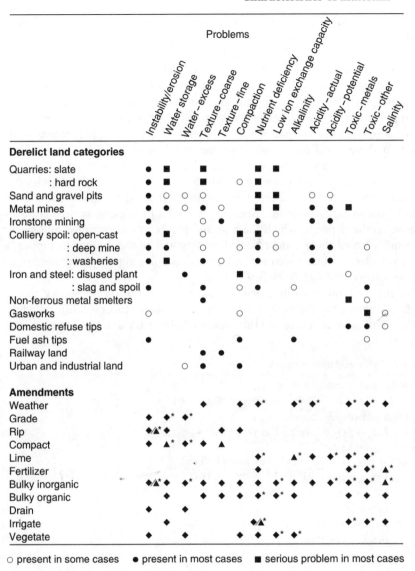

Fig. 5.7 The treatment of substrate characteristics for revegetation (source, Welsh Development Agency, 1993).

on many sites where the materials themselves can be ameliorated to provide a substrate suitable for plant growth. In Fig. 5.7 the ways in which various substrate characteristics can be ameliorated in order to revegetate a site are indicated.

The assessment of contamination has been dealt with in section 4.3 and is relevant to the establishment of plants on toxic materials. Of the toxicities found on degraded sites acidity due to pyrite oxidation (colliery spoil and some metal mine spoils) and high concentrations of heavy metals (metal mines) are frequent problems which have to be overcome during revegetation.

Table 5.4 shows the reactions which take place during pyrite oxidation. Pyritic colliery spoils will continue producing acidity for many years with resultant spoil pHs of 2–3. Vegetation is unable to establish at this low pH and applications of lime unless given in massive amounts are often only effective for short periods. Despite much research into methods of preventing pyrite oxidation there is no one solution to the problem. Table 5.5 suggests some strategies for the treatment of pyritic sites for vegetation establishment. Creation of an oxygen-consuming rhizosphere is a relatively recent technique of which most success has been achieved in the United Kingdom using large applications of sewage sludge. Such applications of organic matter are only effective to the depth to which the material has been incorporated and so for tree establishment there is the risk that rooting will be too shallow to support a fully grown tree and management techniques such as coppicing may need to be used to restrict tree size in such situations.

Metalliferous mine spoils usually contain concentrations of heavy metals which are above the thresholds at which most plants can establish or which restrict plant growth. Direct establishment of plants on untreated mine wastes

Table 5.4 Pyrite oxidation reactions

The overall reaction can be written:
$$FeS_2 + \tfrac{15}{4}O_2 + \tfrac{7}{2}H_2O \rightleftharpoons Fe(OH)_3 + 2SO_4^{2-} + 4H^+ \tag{5.1}$$
if the pH is greater than 2.3, or
$$FeS_2 + \tfrac{15}{8}O_2 + \tfrac{13}{2}Fe^{3+} + \tfrac{17}{4}H_2O \rightleftharpoons \tfrac{15}{2}Fe^{2+} + 2SO_4^{2-} + \tfrac{17}{2}H^+ \tag{5.2}$$
if the pH is less than 2.3.
Intermediate reactions (5.3) to (5.7) have been identified:
$$FeS_{2(s)} + \tfrac{1}{2}O_2 + 2H^+ \rightleftharpoons Fe^{2+} + 2S^\circ{}_{(s)} + H_2O \tag{5.3}$$
$$2S^\circ{}_{(s)} + 3O_2 + 2H_2O \rightleftharpoons 2SO_4^{2-} + 4H^+ \tag{5.4}$$
$$Fe^{2+} + \tfrac{1}{4}O_2 + H^+ \rightleftharpoons Fe^{3+} + \tfrac{1}{2}H_2O \tag{5.5}$$
$$Fe^{3+} + 3H_2O \rightleftharpoons Fe(OH)_{3(s)} + 3H^+ \tag{5.6}$$
$$FeS_{2(s)} + 14Fe^{3+} + 8H_2O \rightleftharpoons 15Fe^{2+} + 2SO_4^{2-} + 16H^+ \tag{5.7}$$

Notes:

$_{(s)}$ Solid

(5.3) Pyrite is oxidized by oxygen. This is a slow reaction, the rate limited by the diffusion of oxygen-releasing ferrous ions and elemental sulphur.

(5.4) Elemental sulphur is oxidized to sulphate ions and acidity, as H^+.

(5.5) Under aerobic conditions, ferrous ions are oxidized to ferric ions. This reaction is catalysed by iron-oxidizing micro-organisms such as *Thiobacillus ferrooxidans*.

(5.6) Above pH 3.5, ferric ions are not stable in water. Ferric hydroxide is formed which precipitates, and pH is lowered further. Ferric hydroxide is very insoluble, so few Fe^{3+} ions are left in solution.

(5.7) Any Fe^{3+} ions remaining in solution are free to oxidize pyrite. Further acidity is generated by this reaction. Since ferric ions are stable below pH 3.5 and are not removed by precipitation, this step is significant in the production of very acid spoil or drainage.

Table 5.5 Strategies for the control of acid generation

Factors limiting pyrite oxidation	Strategies	Treatments
Above pH 4, oxidation is limited by oxygen supply	• Restrict O_2 diffusion	• Compaction • Create O_2-consuming rhizosphere using organic matter, plant roots and aerobic micro-organisms. • Apply 'oxygen-barrier' of soil at least 0.5 m deep
Below pH 4, oxidation is catalysed by bacteria and is not O_2-dependent	• Maintain pH above 4 • Minimize bacterial population/activity	• Apply lime • Maintain pH above 4 • Apply bactericides or inhibitors

using commercially available seed often results in failure. Tailings from abandoned mines are usually very toxic. Metal recovery during ore processing at operating mine sites is becoming so efficient that newly produced non-pyritic tailings at many sites are not as restrictive to plant growth for toxicity reasons as in older, less efficient operations.

Metalliferous mine spoils are nitrogen-deficient also and although the overriding limitation on plant growth may be due to toxic metals, the principles governing nutrient cycling still apply. Metals present affect not only the higher plants but also the microbes governing nitrogen cycling and vegetation establishment can be improved by the addition of fertilizer and organic matter. Two mechanisms seem to operate: one is that at higher nutrient levels there is an increased ability of the plants to cope with stress due to toxicity, and the other is that the chelation of metals by organic matter reduces the availability of these metals for uptake by plants.

Some plants are able to tolerate the high concentrations of metals on metal-rich sites and of these some have been selected and produced commercially specifically for metal mine reclamation. In the United Kingdom selection of such ecotypes has been from Welsh metal mines as follows:

1. *Festuca rubra* cv. Merlin – selected from Trelogan lead/zinc mine, Clwyd. Suitable for neutral and calcareous lead/zinc wastes and dry nutrient-poor materials.
2. *Agrostis capillaris* cv. Goginan – selected from Goginan lead/zinc mine, Dyfed. Suitable for acidic lead/zinc wastes.
3. *Agrostis capillaris* cv. Parys – selected from Parys Mountain, Gwynedd. Suitable for acidic copper wastes.

The advantages of using metal-tolerant species are:

1. They are relatively cheap to establish.
2. It is not necessary to import covering material.

3. It is not necessary to remove contaminated material prior to vegetation establishment.
4. Site disturbance is minimal, reducing the risk of metal dispersal.
5. The vegetation cover will provide some erosion control.

The disadvantages are:

1. If grazing is intended this will need to be carefully controlled so that grazing animals do not ingest large quantities of metals.
2. Run-off will be contaminated with metals and so vegetation establishment alone may not be a suitable treatment near watercourses.
3. Use of the site will need to be restricted so that sensitive members of the population (e.g. children) do not have prolonged contact with spoil material.
4. Erosion may occur on steep slopes if the spoil material vegetated is fine-grained and erosion control measures (such as the use of geotextiles) are not used.

Most degraded land materials will benefit from the application of organic matter in order to establish and promote plant growth, when a reclamation to productive agro-forestry is the aim. The characteristics of some organic materials are given in Table 5.6. Other locally available materials may be appropriate and the principal considerations in selecting materials are:

1. Their carbon : nitrogen ratio.
2. Physical characteristics.
3. Presence of any phytotoxic substances.
4. Presence of any substances which may leach out and cause damage to the wider environment, e.g. ground- or surface water.
5. Aesthetic considerations such as odour and appearance.

In addition to organic amendments inorganic amendments can be used to improve substrate characteristics. Some examples of inorganic amendments are provided in Table 5.7.

Some degraded land substrates can be improved physically to provide better conditions for plants. Examples of physical improvement techniques are provided in Table 5.8. An interesting case is that of Gibson and Looney (1994) who investigated the effect of adding sand to a beach on Perdido Key, Florida. They demonstrated that the vegetation which became established as a result of this practice had to be allowed to equilibrate before any further addition of sand was possible. Uresk and Yamamoto (1986) showed that a dramatic improvement in the biomass of different species grown on bentonite mine spoil could be effected by a range of soil amendments (Fig. 5.8). Lax *et al.* (1994) investigated the potential for the use of mitigating the effects of irrigation of soil with salinized water of the addition of municipal solid waste (MSW). They found significant improvements in the physical characteristics of the soil, with some ameliorative effects coming from improved leaching

Table 5.6   Organic soil amendments (after Coppin and Bradshaw, 1982)

Material	Usual composition (% dry solids)			Organic matter	Usual application rates (dry t/ha)	Special problems or advantages
	N	P	K			
Farmyard manure	0.6–2.5	0.1	0.5	24–50	5–40	Variable
Pig slurry	0.2–4.0	0.1	0.2	3	5–20	High water content, possibly high Cu
Poultry manure, broiler	1.5	0.9–2.5	1.6–2.5	60–80	2–10 }	High levels of ammonia, odours
Poultry manure, battery	2.0–4.0	0.5	0.6	35	2–10	
Sewage sludge, digested	2.0–4.0	0.3–1.5	0.2	45	5–50 }	Possibly toxic metals and pathogens, odours
Sewage sludge, raw	2.4	1.3	0.2	50	5–50	
Mushroom compost	2.8	0.2	0.9	95	5–10	High lime content
Domestic refuse, composted	0.5	0.2	0.3	65	20–70	Contains miscellaneous objects
Brewery sludge, digested	1.5	0.9	0.3		5–20	Uncommon
Peat	0.1	0.005	0.002	50	5–10	Variable, high carbon to nitrogen ratio. Production may cause destruction of wetland habitats
Straw	0.5	0.1	0.8	95	5–10	Decomposition uses soil nitrogen
Sawdust	0.2	0.02	0.15	90	10–30 }	High carbon to nitrogen ratio, requires pulverizing, maturing or composting
Woodchips	0.2	0.02	0.1	90	10–30	
Bark	0.3	0.09	0.7	90	10–30	
Lignite, ground	1	0	0	0		High cation exchange capacity

in the MSW-amended soils, leading to a substantial increase in the fresh weight of a tomato crop grown on the soils in relation to a saline-irrigated but unamended soil.

This type of substrate improvement is becoming increasingly common, but remains very much in the field of the experimental approach, as the local substrate availability and *availability of the amendment* will determine exactly what is applied and when.

5.4   Drainage

Timely and adequate drainage is where the major gains in the improvement of reclamation to agriculture have occurred in the last 25 years. Because of the

Table 5.7   Inorganic amendments

Amendment	Characteristics
Subsoil	An excellent physical amendment, free of weeds. Organic matter and nutrients must be added as required. Variable in texture and must be handled accordingly (see Chapter 3).
Overburden	Like subsoil but likely to be more stony and with less fine-grained material.
Quarry waste	Minerals rejected during quarrying and processing because of wrong quality or size. Ranges from fine-grained material to large size and sometimes with good particle size distribution for plant establishment. Can be used as a substrate for plant growth or as drainage or break layers depending on characteristics. Selection of material important.
Pulverized refuse	Fibrous with inert fragments of solid materials. Contains organic matter as well as inorganic materials. Fertile and water-retentive. Should be incorporated in substrate to be treated to prevent formation of a mat.
Coal washery tailings and slurries	Can have a texture and water-retentive properties of a sandy loam. Texture, pH, pyrite content and presence of flocculating chemicals should be tested for before use.
Pulverized fuel ash (PFA)	Texture of a fine-grained silty soil. Unless weathered, PFA is alkaline, saline and may contain high concentrations of boron which is phytotoxic. Weathered material is excellent for covering coarse skeletal soils.
Dredged silt	Good for amending free-draining materials. If from a marine source can be saline and in industrial areas may contain heavy metals and other contaminants.
Demolition rubble	After crushing produces a suitable substrate for grass establishment. Brick and mortar rubble have a pH above neutral and contain phosphate, potassium, calcium and magnesium.

poor physical state of the soil, water relations will be of paramount importance. Drainage is designed to get water off the site as quickly as possible, to avoid the problems of anaerobiosis associated with the onset of waterlogging. Unfortunately the majority of soils will be *shedding* rather than *infiltrating*. This simply means that the water tends to run off the surface of the soil very quickly during rainfall, as the soil is waterlogged rapidly, rather than draining through it. This leads to the possibility of losses of large amounts of particulate matter, i.e. silts and clays, giving rise to problems of siltation further down large-scale drainage schemes, in ditches and water-bodies, and the loss of large amounts of indigenous and added nutrients. This can lead to serious pollution problems in water-bodies, with high chemical and biological oxygen demand material entering streams, rivers, ponds and lakes. This will stimulate the growth of micro-organisms, which will metabolize the compounds and make use of the nutrients, giving rise to a rapid exhaustion of oxygen, required in the

Table 5.8 Physical improvement techniques (Welsh Development Agency, 1993)

Technique	Application	Equipment	Comment
Crushing	Building rubble, brick, soft rock, shales, slate waste.	Stone crusher, tracked machine and roller, agricultural cultivation equipment.	Demolition and clearance by tracked machines can leave a fine-grained tilth suitable for grass seeding. Repeated passes with rollers and use of stone crushers are required on harder materials. Agricultural cultivation equipment can be used as a secondary treatment to produce a better seedbed.
Compacting	Loose-tipped substrates, e.g. colliery spoil.	Wheeled or tracked machine and roller.	Compaction reduces pore size and so increases water retention. Only free-draining materials should be compacted. Over-compaction can impede root penetration.
Ripping	Compacted substrates, e.g. settled, concreted substrates such as tailings or previously compacted substrates such as former haul roads, car-parks, areas beneath buildings.	Tracked or wheeled machines with deep or shallow tines.	Allows drainage, aeration and root penetration. Best results are achieved if ripping pattern follows fall of ground and/or ripping penetrates underlying free-draining strata. Ripping may bring to the surface buried debris.
Grading	Eroding slopes and uneven ground.	Wheeled or tracked machine with blade.	Can be used to reduce slope to more stable angles, smooth uneven ground and clear debris.
Drainage	Waterlogged and poorly drained materials.	Excavator or specialist drainage machinery.	Suitable outfalls for drainage water are needed. A large number of drainage techniques are available and should be matched to particular site needs.
Addition of coarse-grained material	Fine-textured substrates.	Agricultural cultivation equipment.	Prevents settlement and maintains drainage and aeration.
Addition of fine-grained material	Coarse-textured substrates.	Agricultural cultivation equipment.	Increases water retention.
Covering	Toxic or unsuitable materials.	Dump trucks, bulldozers, excavators.	Details of covering techniques are provided in section 4.4.2.

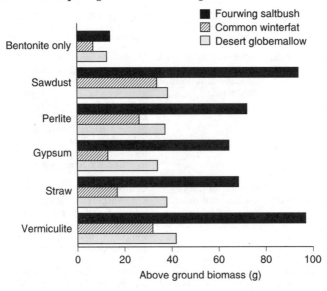

Fig. 5.8 Effects of organic amendment on plant growth on bentonite spoil (redrawn from Uresk and Yamamoto, 1986).

oxidation of these compounds. This lack of oxygen will result in deaths of fish and invertebrates.

This loss of particulate matter and nutrients also reduces the suitability of the soil for plant growth. There are two main types of drainage:

1. **Surface drains and ditches**. These may be excavated at the field edges, and are very common across slopes. They function to take water away from the field as efficiently as possible and prevent shedding into other fields and water features (Fig. 5.9). Ditches often form the major engineered feature of new field boundaries.
2. **Under-drainage**. This involves the provision of cylindrical tunnels under the surface of the soil, which allow infiltrating water to drain away. There are a number of ways in which these drains may be formed (Fig. 5.10). Temporary 'mole' drains are formed by a bullet-shaped metal tine (10–15 cm diameter) being pulled through the soil, at a depth of 30–50 cm. Unfortunately this type of drain does not tend to last long on a disturbed site, through collapse or by becoming blocked. It is more common to install permanent plastic pipes with holes at regular intervals, by digging a ditch, laying the drainpipe, backfilling with aggregate and re-covering with soil. This type of drainage gives a much more satisfactory result, and is both more efficient and longer-lasting.

Depending upon the topography of the site, the drains may be spaced parallel at 10–50 m intervals (tending toward smaller spacing on reclaimed sites), and

Fig. 5.9 Field ditch for drainage on a reinstated mining site.

may be laid in a herringbone pattern down the slope. Both types will eventually feed into a drainage ditch.

Younger (1989) has highlighted the importance of drainage in restoration and reclamation, certainly as regards to agriculture. It was clearly demonstrated that providing an effective drainage scheme had a greater impact on crop yields than soil-handling history, and this could be capitalized upon by nitrogen fertilization and good husbandry.

Ditch   Plastic pipe   Mole drain
         laid in trench   formed by
         backfilled with   bullet tine
         aggregate

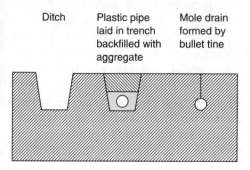

Fig. 5.10 Some types of field drain used in land reclamation and restoration programmes.

5.5   The role of vegetation in reclamation and restoration

The lack of plant nutrients in degraded land substrates was discussed above, and in Chapter 6 the need for plant nutrients to accumulate for succession to proceed and the role of legumes in assisting the build-up of nitrogen will be referred to. It may not be the aim of reclamation to promote ecological succession, but whether the reclamation scheme is intended for intensive use such as playing fields or less intensive use such as a country park, efficient nutrient cycling will be required, and the role of vegetation in this has to be understood.

Derelict land will develop a vegetation cover and efficient nutrient cycling if given long enough. In the absence of legumes, and therefore relying on inputs of nutrients from rainfall, the process of nitrogen accumulation will be slow with development of organic matter being through the accumulation of plant material of wide carbon to nitrogen ratio. Such material is slow to decompose and acts as a 'sink' for any available nitrogen which does enter the system. On reclaimed sites the aim is often to accelerate the process so that vegetation is supplied with its nutrient needs from cycling of nutrients within the soil–plant system or through the addition of fertilizer or both. Although there has been much research into the nutrient cycling of derelict land the questions 'what causes nutrients to build up and cycle on derelict land?' and 'how can this be accelerated during the reclamation process?' are likely to elicit the answer 'it depends', even from the most experienced practitioner or researcher. In Table 5.9 the factors affecting nutrient build-up in reclaimed land are summarized, indicating where the processes of nutrient accumulation may be accelerated or constrained.

The factors listed in Table 5.9 can be manipulated to accelerate the accumulation of nutrients or more importantly to establish a nutrient cycle which provides nutrients at a speed and level which meets the requirements of the established vegetation and the use of the site. Very often in the reclamation of derelict land the vegetation appropriate to the proposed land-use is established immediately after earthworks have been completed. By contrast, in the reclamation of open-cast mined land with afforestation and with agriculture, a pioneer crop of a short duration is established as part of a programme of soil improvement before the final crop or vegetation is established. Pioneer crops can aid the accumulation and cycling of nutrients by providing organic matter, improved soil structure and nitrogen (if, for example, legumes are used). In addition such crops can provide shelter for exposure intolerant crops. Annual grasses such as 'Westerwolds' ryegrass can be used to protect trees and shrubs sown directly into derelict land substrates. In some schemes cereals or shrubby legumes such as lucerne have been used to provide organic matter and nitrogen and are cultivated into the substrate at the end of the growing season. A difficulty with this approach on many derelict land substrates is that they are often not easy to cultivate and cultivation may bring unwanted materials to the surface. Short-lived shrubs such as the legumes

Table 5.9 Factors affecting nutrient accumulation on reclaimed derelict land

Factor	Effect
Herbaceous swards Inorganic fertilizer application	**Nitrogen**: nitrogen applied will be immediately available to establishing and growing plants but will be easily leached. In order for growth of plants to be satisfactory and organic matter to be decomposed, frequent nitrogen applications have to be made. The consequences of not applying frequent nitrogen are more extreme on derelict land materials than soils where nitrogen is mineralized from soil organic matter. Infrequent application of nitrogen fertilizer on derelict land substrates will result in the immobilization of nitrogen in plant material of a wide carbon to nitrogen ratio. **Phosphate**: phosphate applied may be immediately available or more slowly depending on the type of fertilizer used. Although some derelict land materials such as colliery spoil can 'fix' phosphate making it unavailable to plants, phosphate is not as readily leached from derelict land materials and need not be applied as frequently as nitrogen. Unless productive vegetation is needed, phosphate applications can be applied at three-year intervals or greater. Where maximizing of leguminous nitrogen fixation is required, more frequent phosphate applications are likely to be needed. Because of the lack of leaching losses of phosphate, accumulation of phosphate in the plant–soil system occurs more readily than that of nitrogen. **Potassium**: potassium is in moderate supply in derelict land materials containing clay minerals such as colliery spoil but in short supply in materials such as china clay waste and sandy substrates. Grasses will take potassium up in 'luxury' proportions if given the opportunity and such potassium accumulates in organic matter and then binds to clay minerals. However, potassium is easily leached from sandy materials and repeated supply may be needed.
Organic fertilizer application	The properties of organic fertilizers have been summarised in Table 5.6. In general terms application of organic fertilizers results in slower release of nutrients, less leaching of nutrients and greater accumulation of nitrogen than inorganic fertilizers for an equivalent amount of nitrogen input. However, care has to be taken that nitrogen in particular is not lost through volatilization as ammonia from some materials (e.g. poultry manure) or is immobilized in material of a wide carbon to nitrogen ratio.
Legumes	Legumes my require some nitrogen fertilizer to establish and require phosphate to fix nitrogen adequately. A healthy legume grass sward will if properly managed both supply nitrogen to a sward on a more consistent basis than fertilizer applications and accumulate organic nitrogen in the soil particularly in grazed or mown systems where some of the nitrogen is removed and not fully returned (Figs 5.11 and 5.12).

Table 5.9   (contd.)

Factor	Effect
Defoliation	Grazing and cutting will remove or prevent the accumulation of standing dead vegetation of a wide carbon to nitrogen ratio. Grazing will return some of the nitrogen removed to the soil in an organic form which will be mineralized to provide nitrogen available to plant roots. Grazing therefore allows the accumulation of organic matter in a form which will be rapidly recyclable providing the grazing and any necessary nutrient inputs are properly managed. Overgrazing is likely to lead to swards which are under stress with little accumulation of organic matter. Cutting will remove nutrients and prevent the rapid accumulation of organic matter unless the cuttings are returned. If the cuttings are returned, their rate of decomposition will depend on their carbon to nitrogen ratio and the amount of nitrogen available to reduce the carbon to nitrogen ratio if this is wide.
Trees and shrubs	There is not much data on the accumulation of organic matter under trees on derelict land materials but where trees are growing well, surface accumulation of organic matter of a wide carbon to nitrogen ratio is likely to occur derived from leaf fall. On slopes, however, these leaves and resultant organic matter may accumulate at the base of the slope. In the long term, soil organic matter will be derived from the death and sloughing off of root material. Where the trees are leguminous, recycling of nodules will also contribute to soil organic matter.

lupin, gorse and broom have been used to protect young seedlings and provide additional nitrogen. Pioneer trees such as alder and birch can fulfil the additional role of providing a woodland environment suitable for the growth of canopy species such as oak, ash and beech.

The effectiveness of any vegetation in fulfilling the role of nutrient accumulation and cycling is dependent on the ability of the species to survive and grow under the conditions on site. There is a wide range of cultivated varieties of grasses available for reclamation to non-agricultural uses. Most of these grasses have been selected and bred for amenity uses such as sports pitches and golf courses although some have been selected for tolerance of high levels of metals, and cultivars of flattened meadow grass have been introduced into the United Kingdom because of their success in revegetation of colliery spoil in North America. Despite the large number of cultivated varieties available, some grasses are not able to tolerate a wide range of substrate conditions (Fig. 5.13). Perennial ryegrass is such an example, being only suitable for fertile conditions where there are no extremes of pH or moisture. In comparison, red fescue can tolerate a wide range of conditions. This is partly explained by the fact that it has a large number of naturally occurring ecotypes tolerant of different conditions from which cultivars have been selected. However, exhaustive research on the suitability of grass cultivars for the revegetation of colliery spoil (Derelict Land Reclamation Research Unit, 1982) has demonstrated that single cultivars of red fescue are able to tolerate a wide range of conditions and in particular can withstand falls in nutrient availability and then respond when

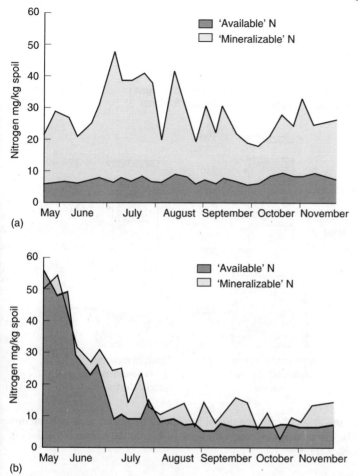

Fig. 5.11 Fertilizer and legume-based nitrogen supply through the season in grass swards on colliery spoil. (a) Grass/legume sward, no nitrogen fertilizer additions; (b) grass sward maintained by fertilizer additions in early May. 'Available' nitrogen is that immediately available in a mineral form (NH_4^+, NO_3 or NO_2^-). 'Mineralizable' nitrogen is that capable of being mineralized from spoil organic matter and is an indication of spoil fertility (source, Palmer, 1984).

nutrients become more available. Perennial ryegrass is likely to become moribund under the same set of conditions.

5.5.1   Engineering function

Vegetation performs a range of engineering functions at different stages in the land reclamation process (Fig. 5.14). For established vegetation these functions can be divided into mechanical and hydrological functions (Fig. 5.15). The

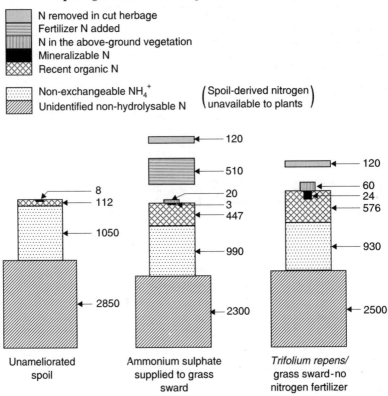

Fig. 5.12 The accumulation of nitrogen on colliery spoil seven years after revegetation (after Palmer, 1984). The columns represent the relative sizes of the nitrogen fractions in kg N/ha in 150 mm of spoil.

effects can be both beneficial and adverse, depending on the substrate type, type of vegetation and climate. In Table 5.10 the conditions under which a vegetation cover can be beneficial or adverse are summarized in relation to land reclamation. The properties of vegetation are seasonal in nature and vary with vegetation type and plant species and age. The influence and interaction of such vegetation properties between themselves and engineering function are summarized in Fig. 5.16.

5.5.2 Decontamination

It has long been known that some plants are able to tolerate higher concentrations of metals in soils than others. Plant strategies for achieving tolerance vary but can broadly be divided into exclusion and accumulation. Excluders are those plants where metal concentrations in the shoot are maintained at a constant level regardless of the metal content of the soil. Accumulators are those

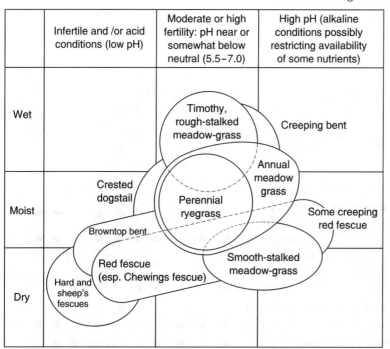

	Infertile and /or acid conditions (low pH)	Moderate or high fertility: pH near or somewhat below neutral (5.5–7.0)	High pH (alkaline conditions possibly restricting availability of some nutrients)
Wet		Timothy, rough-stalked meadow-grass	Creeping bent
Moist	Crested dogstail Browntop bent	Perennial ryegrass	Annual meadow grass Some creeping red fescue
Dry	Hard and sheep's fescues	Red fescue (esp. Chewings fescue)	Smooth-stalked meadow-grass

Fig. 5.13   Soil preferences of some common turf grasses (after Shildrick, 1984).

plants in which metals are concentrated in the above-ground plant parts from low or high soil levels (Baker, 1981). Indicators are those species where plant concentrations of metals reflect those in the soil. Different ecotypes may exhibit different mechanisms at the same concentration of metal in the soil and the same species may exhibit different mechanisms of tolerance at different concentrations of metals (Fig. 5.17). Interest has developed in the ability of accumulators to remove metals selectively from soils as a means of decontaminating soil. Although no large-scale decontaminations using such strategies have taken place trials have indicated that the capacity for removal is quite considerable (Baker *et al.*, 1994).

Plants may assist in two ways in the degradation of persistant organic chemicals in the soil: they may take up and partly or fully metabolize certain organic chemicals, rendering them less toxic (Bell and Failey, 1991), or they may support chemical-degrading micro-organisms in the soil. Existing methods of *in situ* bioremediation use bacteria or fungi that are able to live on and thus degrade certain toxic organic chemicals, but their performance in unplanted soil is often unsatisfactory (see Chapter 4). Plant roots are known to support very high populations of, and metabolic activity by, micro-organisms in their vicinity (the rhizosphere). Plants have also been shown to increase the survival of bacteria or symbiotic fungi that have the ability to degrade toxic chemicals (Anderson *et al.*, 1993; Walton *et al.*, 1994). However, phytoremediation has

Applications	Engineering situations										
	Mining and reclamation	Construction sites	Waste disposal and pubic health	Waterways	Land drainage	Reservoirs and dams	Coastal and shoreline protection	Buildings	Recreation	Pipelines	Site appraisal
Slope stabilization – embankments and cuttings	●	●				●	●				
– cliffs and rockfaces	●						●				
Water erosion control – rainfall and overland flow	●	●	●		●	●	●			●	●
– gully erosion	●	●		●	●	●	●			●	●
Water course and shoreline protection – continuous flow channels				●	●						
– discontinuous flow channels	●		●	●	●		●	●			
– large water bodies (shorelines)						●	●				
Wind erosion control	●	●	●				●			●	●
Vegetation barriers – shelter	●		●	●			●	●	●		
– noise reduction			●					●			
Surface protection and trafficability		●			●	●	●		●	●	●
Control of run-off in small catchments	●	●	●	●	●				●		
Plants as indicators	●		●	●		●				●	●

Fig. 5.14 The engineering role of vegetation in land reclamation applications (after Coppin and Richards, 1990).

also a number of possible limitations, in particular the limited depth of contamination that can be treated, the potential for accumulation of toxic non-degradable metabolites, and the limited availability of many organic and inorganic contaminants for plant uptake and microbial breakdown.

Using plants in this way will require a change in philosophy with respect to the reclamation and development of sites as the process of decontamination by such means is likely to take a long time. Other considerations are indicated in Table 5.11.

5.5.3 Visual impact

The role of vegetation in the visual impact of a scheme cannot be completely divorced from other considerations such as its engineering role or the ecology

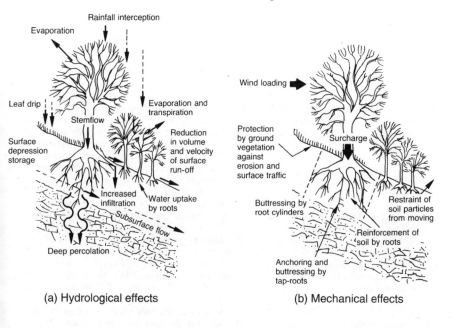

Rainfall interception

Evaporation

Leaf drip

Stemflow

Evaporation and
transpiration

Surface
depression
storage

Reduction
in volume
and velocity
of surface
run-off

Increased
infiltration

Water uptake
by roots

Subsurface flow

Deep percolation

(a) Hydrological effects

Wind loading

Protection
by ground
vegetation
against
erosion and
surface traffic

Surcharge

Buttressing by
root cylinders

Restraint of
soil particles
from moving

Reinforcement of
soil by roots

Anchoring and
buttressing by
tap-roots

(b) Mechanical effects

Fig. 5.15 The hydrological and mechanical effects of vegetation (source, Coppin and Richards, 1990).

of the site. However, vegetation can be used to good visual effect both in temporary and permanent ways. The overall effect of vegetation may be one of giving the site some maturity or blending with the surroundings, but there are other more subtle effects (Table 5.12).

5.6 Agricultural reclamation

The exact procedure for a successful reclamation to agriculture will be dependent upon local conditions and practice, as related to regions, nations and continents. In the United Kingdom the following activities commonly take place (I. Carolan, 1995, personal communication).

5.6.1 Cultivation

There are two principal objectives of cultivation when the profiles have been reinstated:

Table 5.10   Beneficial and adverse effects of vegetation in land reclamation (adapted from Coppin and Richards, 1990)

Mechanism	Effect[a]	Implication for land reclamation
Hydrological		
Foliage intercepts rainfall causing:		
1. Absorptive and evaporative losses reducing rainfall available for infiltration.	B	Prevention of soils becoming waterlogged, yet retention of water for use by plants and binding of soil. Important on clayey or silty substrates such as colliery spoil which can be waterlogged in the winter and droughty in the summer. Lower infiltration reduces soil weight which is an important factor on slopes where soils may be potentially unstable particularly if they have been recently placed as in a reclamation scheme.
2. Reduction in kinetic energy of raindrops and thus erosivity.	B	A very important function on many reclamation schemes on all soil types. One of the principal engineering reasons why a quick cover of vegetation is desirable in reclamation.
3. Increase in drop size through leaf drip thus increasing local rainfall intensity.	A	Depends on the vegetation type. For grasses and most herbaceous swards in the United Kingdom this effect is not important. However, for trees and shrubs particularly where there is no ground cover vegetation this effect can cause increased erosion in the rows between plants.
Stems and leaves interact with flow at the ground surface, resulting in:		
1. Higher interception storage and higher volume of water for infiltration.	A/B	Whether the effect is beneficial or not depends on whether the substrate is near to waterlogging and infiltration is beneficial. As water which does not infiltrate is likely to run off, a process which easily leads to erosion of reclamation schemes, increased infiltration is usually beneficial.
2. Greater roughness on the flow of air and water, reducing its velocity.	B	Reduces erosion.
3. Tussocky vegetation may give high localized drag, concentrating flow and increased velocity.	A	Heterogeneity of substrate and poor or inappropriate sowing and fertilizing regimes or plant material may give rise to uneven vegetation cover leading to erosion by this mechanism.
Roots permeate the soil leading to:		
1. Opening up of the surface and increased infiltration.	A	Increases moisture content over unvegetated areas, although increased evapotranspiration may offset this. As with other mechanisms increasing infiltration, its effect is soil-dependent.
2. Extraction of moisture which is lost to the atmosphere in transpiration, lowering pore-water pressure and increasing soil suction, both increasing soil strength.	B	Poor soil strength is a feature of many derelict land substrates and materials used in covering in reclamation schemes. Factors which increase soil strength are therefore very beneficial.
3. Accentuation of desiccation cracks, leading to higher infiltration.	A	Increases soil weight and potential for waterlogging as with other mechanisms which increase infiltration.

Table 5.10   (contd.)

Mechanism	Effect[a]	Implication for land reclamation
Mechanical		
Roots bind soil particles and permeate the soil, resulting in:		
1. Restraint of soil movement reducing erodibility.	B	Increases in the cohesion of the soil are very beneficial in reclamation schemes where substrates may be structurally weak.
2. Increase in shear strength through a matrix of tensile fibres.	B	As above.
3. Network of surface fibres creates a tensile mat effect restraining underlying strata.	B	As above.
Roots penetrate deep strata giving:		
1. Anchorage into firm strata, bonding soil mantle to stable subsoil or bedrock.	B	Beneficial where new materials have been placed over stable strata.
2. Support to upslope soil cover through buttressing and arching.	B	Beneficial on slopes. Trees and other deep-rooted vegetation will maintain some stability of slopes if they can be left *in situ* during a reclamation scheme.
Tall growth of trees so that:		
1. Weight may surcharge the slope increasing normal and down-slope force components.	A	Can cause slippage of slope materials. Can be of concern on loose tipped materials such as colliery spoil where the initial soil binding and stabilizing effects of vegetation are superseded by surcharging resulting in slippage. Management of trees by a coppicing regime will retain soil-binding properties of the vegetation and reduce risk of surcharging.
2. When exposed to wind, dynamic forces are transmitted into the ground.	A	Leads to windthrow which is more prevalent in tree planting on spoil heaps than natural soils.
Stems and leaves cover the ground surface so that:		
1. Impact of traffic is absorbed protecting soil surface from damage.	B	Important not only for vehicular traffic but for humans and animals. On some spoil materials grazing has to be restricted to the drier parts of the year even where there is a good vegetation cover because of damage to the soil in wet conditions through trampling.
2. Foliage is flattened in high velocity flows, covering the soil surface and providing protection against erosive flows.	B	Beneficial particularly where surface has recently been disturbed or placed.

[a]A = adverse, B = beneficial.

Vegetation		Vegetation properties										
Effect on	Influence	Ground cover (%)	Height	Leaf shape and length	Stem/leaf density	Stem/leaf robustness	Stem/leaf flexibility	Root depth	Root density	Root strength	Annual growth cycle	Weight
Surface competence	Soil detachment	•	•	•	•						•	
	Mechanical strength	•	•		•	•			•	•	•	
	Insulation	•			•						•	
	Retarding/arresting			•		•	•	•				
	Erosion	•			•						•	
Surface water regime	Rainfall interception	•		•	•							
	Overland flow/run-off	•			•							
	Infiltration				•				•	•		
	Sub-surface drainage								•	•		
	Surface drag	•	•	•	•		•				•	
Soil water	Evapotranspiration			•	•				•		•	
	Soil moisture depletion leading to increased soil suction, reduced pore-water and soil weight							•			•	
Properties of soil mass	Root reinforcement							•	•	•	•	
	Anchorage/restraint							•	•	•		
	Arching/buttressing							•		•		
	Surface mat/net								•	•	•	
	Surcharge		•									•
	Windthrow		•		•	•		•		•	•	•
	Root wedging							•	•			
Air flow	Surface drag	•	•	•			•				•	
	Flow deflection	•		•		•	•				•	
	Noise attenuation	•	•	•							•	
	Suspended particulates	•		•							•	

Fig. 5.16 Vegetation properties and their engineering significance (source, Coppin and Richards, 1990).

1. Disruption of compacted horizons, particularly between top- and subsoil layers. This is done in order to prevent the formation of iron-pans by encouraging the development of drainage channels.
2. Production of macro-aggregate crumb structure, in order to provide an adequate seed-bed.

In the first instance this will be carried out by means of pulling winged tines behind a heavy tractor, followed by discing to produce a tilth ready to receive the initial seed mix.

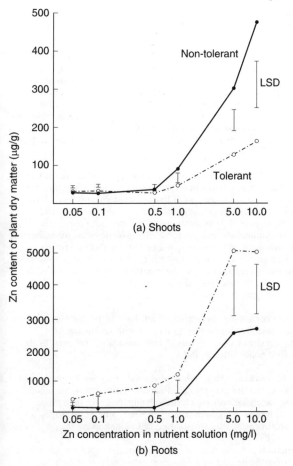

Fig. 5.17 The zinc content of shoots and roots of tolerant and non-tolerant populations of *Silene maritima* growth in solution culture at a range of zinc levels
(LSD = least significant difference) (source, Baker, 1981).

5.6.2 Boundaries

The construction of clear and robust boundaries is important to delimit the
ownership of fields, where more than one previous owner is taking the land
back into ownership, to ensure that livestock does not enter fields undergoing
sensitive or dangerous treatment, or where there is risk of damage to newly
reinstated soils, or where new vegetation is being established.

These boundaries are formed as hedgerows, stonewalls, fences or hedge-on-
bank. Hedgerows usually have a large proportion of quickthorn, which will be
used to deter livestock, and which grows quickly. Increasingly, the reinstate-

Table 5.11 Considerations in using plants to decontaminate soils

Factor	Comment
Choice of plant	Plants will need to be matched to the contaminants in the soil which need removal/destruction. Where removal is concerned existing tolerant ecotypes will have developed tolerances matched to the conditions under which they are growing and may not be tolerant of contaminants with which they have not come into contact. Even if plants are tolerant of contaminants they have not experienced they may have a mechanism of tolerance, such as exclusion, which does not lend itself to decontamination. Where a wide range of contaminants is found and/or the site is very heterogeneous, a range of species may need using.
Vegetation management	Management will need to ensure that the appropriate decontaminating species are maintained under optimum conditions for decontamination. Where decontamination is successful but soil contaminant levels have not reached the desired target, aggressive species able to withstand conditions in the partially decontaminated soil but not themselves assisting decontamination may invade at the expense of decontamination species. Management will need to avoid or remedy this eventuality. Where the decontamination mechanism is concentration of contaminants in the plant material, then for decontamination to be effective this material must be removed from site by cutting. Such removal will also remove nutrients and necessitate higher nutrient inputs than otherwise.
Disposal of material	Plant material containing contaminants will have to be disposed of. This will be bulky and therefore costly to remove without treatment. Composting on site or other means of bulk reduction may need to be adopted before removal to landfill.
Programming	Only parts of a site may be suitable for decontamination by plants and development of the site may therefore need to take place in phases to allow for the timescale over which decontamination will take place. Developers and reclaiming authorities will need to be prepared for long periods of treatment of sites by plant decontamination means.
Depth of treatment	Decontamination is likely only to take place in the rooting zone. Where deeper treatment is needed by plants then excavation or cultivation of soil may be needed to achieve this.

ment of hedgerows is being used as an opportunity to re-establish species-rich linear features to the landscape.

5.6.3 Field drainage

Once the boundaries have been established and ditches formed, extensive under-drainage is installed, which feeds directly into the ditches. On the most severely disturbed sites this will be achieved by laying perforated plastic pipes, as traditional mole drains formed by a bullet-shaped tine are prone to collapse because of the poor structural stability of the soils.

Table 5.12   The visual effect of vegetation in reclamation

Factor	Comment
Integration	Vegetation can integrate a site into its surroundings by masking of substrate colour or dereliction and by softening or hardening features such as landform or boundaries. These effects not only blend the site into the surroundings but can give the site a feel of maturity and naturalness an unvegetated site would not have.
Spatial definition	Vegetation, particularly 'bulky' vegetation such as trees or shrubs, can be used to create spatial structure within a site. This may be achieved by dividing the site into areas defined by bulky vegetation but themselves having vegetation of different colour or texture. Such spatial definition may not only make the site more visually attractive but may provide shelter and also give it an identity and character, making the user feel good about being on the site.
Screening	Vegetation may also be used on site to screen other land-uses, making the site more attractive and contributing to the integration of the site into the surroundings.
Accentuation	Although not often practised, both vegetation and landform modelling can be used to accentuate features of a site so that the site stands out or 'makes a statement'. At its simplest this can be use of, for example, a stark and tall avenue of trees as is sometimes practised in France. The other extreme is to use landform and vegetation in 'earth sculpture', making the site visually prominent.

5.6.4   Field water supply

Livestock drinking troughs are placed in all fields intended for agricultural use. These are connected to a mains water supply. The position of the trough is sometimes changed to minimize the risks of erosion brought about by heavy trafficking from cattle and sheep.

5.6.5   Access

New farm access roads are constructed to serve fields, farms and outbuildings. These may be made up in tarmacadam in some cases, gravelled, or of compacted soil.

5.6.6   Woods and shelter belts

New woodlands are planted in corners and as shelter belts on particularly exposed ridges and hills. Species such as willow (resistant to waterlogging), birch (a resilient pioneer species) and alder (capable of fixing its own nitrogen) are employed, although with great care other species may be used. On many sites one of the major problems in establishing trees is their inadequate mulching and lack of aftercare – in particular there are few attempts made to irrigate new plantings. Although this is costly, to persist in ignoring this fundamental requirement of plant establishment will lead to continued failure in tree areas. This is a particular problem on free-draining materials and ridges. Also many planting schemes are rather unnatural in appearance, being planted in rigid rows. Such areas will never blend into the landscape, and militate against the structural diversity of a site. Refer to section 5.7 for further details.

5.6.7 Non-agricultural features

Increasingly in farmland reclamation there is a diversity of use planned and implemented, including wildlife areas, species-rich meadows, wetlands and public access and amenity use. These will be considered more fully in Chapter 6.

5.6.8 Agricultural rehabilitation

In the first year of reclamation a simple grass/clover ley is sown in order to protect the soil from erosion, and to improve soil structure. This is followed by grazing by sheep, and later cattle, well after the initial five-year aftercare period. Crops such as barley and wheat are not normally grown until the aftercare period, as the soil is still too wet in winter and too dry in summer. There has been considerable success in developing agricultural restoration techniques at the University of Newcastle, resulting in yields of some crops close to (and in some cases exceeding) control values (Younger, 1989; Davis et al., 1992).

5.6.9 Fertilizer and organic inputs

One of the most important aspects of the management of agricultural reclamation is that of maintaining an adequate supply of nutrients to the crops. Nutrient supply is achieved by top dressing with inorganic fertilizers, or incorporation of organic residues.

5.6.10 Management of soil structure

In the not too distant past management of soil structure was a simple matter of reducing the bulk density of units of soils with a resolution no smaller than a square metre. Soil bulk density decreases were accomplished by ripping the soil with a winged tine behind a tractor, which resulted in the large blocks of soil being broken into slightly smaller blocks, with large voids in between, giving an apparently greatly reduced soil bulk density. Such physical treatment does nothing, however, for the micro- and macro-aggregate stability which is essential for the correct drainage and gaseous exchange function of the soil. In order for the fine structure to be enhanced it is necessary to concentrate on the biology of the system, which has a large role to play in the establishment and maintenance of soil structure at this scale (cf. Chapter 1). There must be a combination of two approaches:

1. Manipulation of soil physical and chemical conditions to encourage the initiation and establishment of active soil biological communities.
2. Introduction of selected components of the soil microbial community.

Stewart and Scullion (1989) have stressed the importance of managing the biology of all man-made soils, and there has been a significant effort to investigate this approach carried out at the British Opencast Coal sponsored Bryngwyn Farm project under the auspices of the University of Wales, Aberystwyth (Scullion, 1994). One of the focuses of research was the management and effects of earthworm populations on soil conditions and fertility. There was clear benefit demonstrated by introducing earthworms to reinstated soils, as shown by improvements in pore space and aggregate stability (Fig. 5.18). Scullion and co-workers have also demonstrated that as the intensity of cultivation increased, so the stability of large crumbs decreased, linked to the disturbance of the slowly establishing, fragile, earthworm community. They found that the number of worms per unit area could be increased by increasing the frequency of cutting of grass swards established on the site, from 34 worms

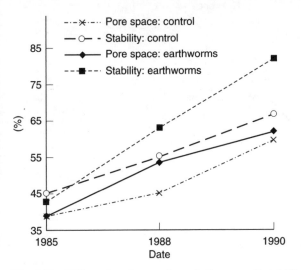

Fig. 5.18   Effect of earthworm inoculation on soil stability and pore space in reinstated soils (redrawn from Scullion, 1994).

per square metre on an uncut ward through $50\,m^{-2}$ if cut once every two months, to $70\,m^{-2}$ if cut once a month.

Edgerton *et al.* (1995) investigated the relationship between the microbial community size and the stability of 4.5 mm aggregates on open-cast coal sites. She discovered a log/linear relationship between the biomass and stability (Fig. 5.19), which suggested two things:

1. The development of stability was linked to the development of the soil microbial community.
2. Enhancement of the community may lead to an increase in the structural stability in soils subject to this treatment.

So how might this latter be carried out? Rogers and Burns (1994) demonstrated that the addition of a nitrogen-fixing cyanophyte *Nostoc muscorum* to a poorly structured silt loam not only improved the nitrogen content of the soil, but also improved structural stability by 18 per cent and increased carbon and potassium after 300 days' incubation at 20°C, in comparison with a control (Fig. 5.20). There were also significant increases in the numbers of bacteria, fungi and actinomycetes, and the rate of seedling emergence. The micro-organism was applied at a rate of 4.04×10^5 cells per gram, equivalent to 5 kg cells per hectare (dry weight of cells). The application of this type of biotechnological method holds out great potential, if it can be applied on a wide enough scale.

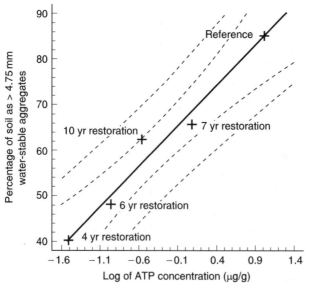

Fig. 5.19 Relationship between microbial biomass (as indicated by adenosine triphosphate concentration) and the stability of > 4.75 mm aggregates (from Edgerton *et al.*, 1995).

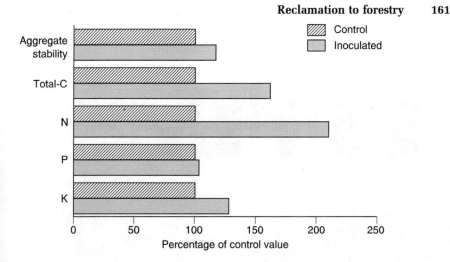

Fig. 5.20 Effect of the addition of *Nostoc muscorum* on soil physicochemical parameters in relation to a control (redrawn from Rogers and Burns, 1994).

5.7 Reclamation to forestry

As in the case of restoration to agriculture after open-cast coal-mining there is a statutory period of five years of aftercare. Unfortunately this is insufficient time to establish plantations properly, and a longer period of at least ten years has been suggested (Moffat and McNeill, 1994). There are significant differences in the success of trees planted on restored sites as compared to undisturbed areas, as demonstrated by Scullion (1994), as shown in Fig. 5.21.

Cultivation of soils and substrates is similar for agricultural reclamation, but because of location and site histories there is a greater likelihood that there will be no topsoil available for planting. In many cases the primary substrate will be mine spoil, i.e. overburden. The type of material that is being planted into will have a major effect on tree survival (Fig. 5.22).

5.7.1 Establishment of vegetation

The primary concern in reclamation to forestry is the establishment and growth of trees for cropping. The species to be selected is probably the principal consideration and must be closely related to the prevailing conditions (Moffat and McNeill, 1994). Due consideration must be given to the soil type, pH, exposure and atmospheric pollution. Most sites for reclamation will have some sort of nutritional shortfall, but when topsoil is being used a wider range of species may be contemplated.

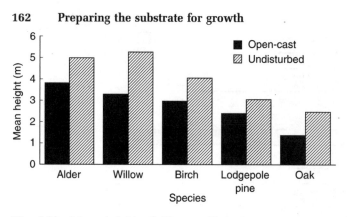

Fig. 5.21 Mean height of 10-year-old trees on open-cast and undisturbed sites (redrawn from Scullion, 1994).

The tree material is available in a variety of forms, as shown in Table 5.13. Choice of stock will depend on the type of substrate, the size of site and the local availability of the stock. It is important that stock of known provenance is used, to avoid problems with poor growth and predation in plantations. They should all be planted when dormant, usually during the late winter/spring in temperate and sub-arctic regions but when the ground is not frozen, and the dry season, if applicable, in the tropics. In most cases in the United Kingdom trees will be spaced at 2 m intervals (Moffat and McNeill, 1994).

The planted stock may need to be protected from animals which will eat leaves, burrow around roots, and strip bark. On a large scale (greater than 2 ha) fencing is employed, but on smaller sites trees will need to be protected

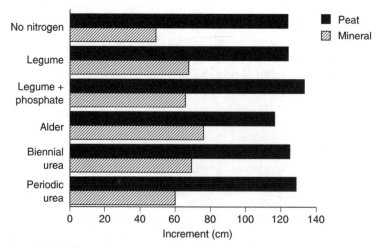

Fig. 5.22 Effect of treatment and substrate type on growth of Sitka spruce over a five-year period (from Moffat and McNeill, 1994).

Table 5.13 Types of stock available for tree establishment

Type	Description	Establishment method	Comments
Transplants	Small plants less than 1.2 m in height, up to 4 years old	Planted bare rooted in notches	Most common type of stock in large planting schemes
Undercuts	Similar to transplants, but roots trimmed to improve rooting pattern and to control growth of shoot	Planted bare rooted in notches	Tends to have a higher rate of survival and establishment, but more expensive
Container grown	Plants grown from seed or cuttings in a pot until ready for use. Smaller plants are usually preferred	Must be planted in a suitably sized hole, cannot be satisfactorily notch-planted	Expensive in relation to bare rooted stock, but has the highest rate of survival and establishment. Particularly useful for establishing sensitive species
Direct seeding	Seeds harvested from natural areas under controlled conditions, or from plantations. Provenance (genetic origin) particularly important	Seeds sown by means of tractor-towed rig, or by hand	Rates of successful establishment not as high as was hoped, in adverse substrates. Low cost in relation to other systems on a per plant basis but low success rates tend to remove this advantage

individually. They will also need to be protected individually when planted amongst established plants on a large area where it would be impossible to exclude rabbits, squirrels and voles. Individual protection can be a simple 'rabbit guard' which is a strip of rigid paper or plastic forming a corkscrew in the shape of a tube, which is easily fitted to the stem of the plant. The guard then expands as the plant grows, but affords little protection against deer and other large mammals. In recent years the 'tree tube' has been developed by the Forestry Commission, which is a tube made of a translucent plastic, fitting over the stem and down into the soil. The tube not only protects the plant against animal damage, but also acts to stabilize and improve the microclimate that the plant is subject to.

In some cases it may be desirable to establish a ground cover of some sort, particularly where there is a need to improve soil conditions with regard to nutrient availability and structural stability (cf. section 5.5). There is a balance to be struck between the growth of the trees and the establishment of the herbaceous layer. The herbaceous layer may compete directly with the trees for water and nutrients, which is problematical when the trees are young, as they take longer to overcome the trauma of planting in the new environment. Grass and legume mixtures are usually employed in this regard: legumes are particularly useful for fixing nitrogen, the grasses for cover and rooting vigour. Most grasses and legumes can be introduced as seed, which may be sown bare, or in a hydroseeding mulch mixture (Fig. 5.23).

Fig. 5.23 Hydro-seeding operations.

5.7.2 Aftercare

There will be losses occurring in the first few years, and when the stocking density falls below 80 per cent then replacement plantings should be made. It may be appropriate in some cases, however, to allow open spaces to develop.

Long-term analysis of soil conditions and foliar samples will give an early indication of the need for fertilization, or other amendment. As in the case of agricultural reclamation, either organic or inorganic amendments can be made. Fertilizers may be added early on, provided fertilization is carried out judiciously, to enable the establishment of a vigorous rooting system. Inorganic fertilizers are usually applied by hand, as access by vehicles is impractical.

Organic fertilizers, such as sewage sludge, are the addition of preference on most sites, and results are generally very good. Sewage sludge helps with the growth of the trees and will improve the ground cover. Unfortunately the cost of transport may make such an application prohibitive. There is evidence that the use of organic amendments can improve the establishment of invertebrates. Simmonds *et al.* (1994) showed that leaf litter depth and cover, and vegetation density, were positively correlated to recolonization by spiders on a rehabilitated bauxite mine.

5.8 Conclusions

The successful establishment of agro-forestry activities on formerly degraded land has been thoroughly investigated in the past 30 years, and significant progress has been made, such that sites restored after major disruption are now self-sustaining in terms of economics. What remains uncertain is taking this to the next phase, that of establishment of self-sustaining systems, which is the subject of the next chapter.

References

ANDERSON T.A., GUTHRIE E.A. and WALTON, B.T. (1993). Bioremediation in the rhizosphere. *Environ. Sci. Technol.*, **27**(13), 2630–2636.

BAKER, A.J.M. (1981). Accumulators and excluders – strategies in the response of plants to heavy metals. *J. Plant Nutrition*, **3**(1–4), 643–654.

BAKER, A.J.M., MCGRATH, S.P., SIDOLI, C.M.D. and REEVES, R.D. (1994). The possibility of *in situ* heavy metal decontamination of polluted soils using crops of metal-accumulating species. *Resources, Conserv. Recycling*, **11**, 41–49.

BELL, R.M. and FAILEY, R.A. (1991). Plant uptake of organic pollutants. In Jones, K.C. (ed.) *Organic contaminants in the environment – environmental pathways and effects*, pp. 189–206. Elsevier Applied Science, London.

BRADSHAW, A.D. and CHADWICK, M.J. (1980). *The restoration of land*. Blackwell, Oxford.

COPPIN, N.J. and BRADSHAW, A.D. (1982). *Quarry reclamation*. Mining Journal Books, London.

COPPIN, N.J. and RICHARDS, I.G. (1990). *The use of vegetation in civil engineering*. Butterworths, London.

DAVIS, R., YOUNGER, A. and CHAPMAN, R. (1992). Water availability in a restored soil. *Soil Use Man.*, **8**(2), 67–73.

DERELICT LAND RECLAMATION RESEARCH UNIT (1982). *The establishment, maintenance and management of vegetation on colliery spoil sites*. University of York, York.

EDGERTON, D.L., HARRIS, J.A., BIRCH, P. and BULLOCK, P. (1995). Linear relationship between aggregate stability and microbial biomass in three restored soils. *Soil Biol. Biochem.*, **27**(11), 1499–1501.

GIBSON, D.J. and LOONEY, P.B. (1994). Vegetation colonization of dredge spoil on Perdido Key, Florida. *J. Coastal Res.*, **10**(1), 133–143.

GILBERT, O. (1989). *The ecology of urban habitats*. Chapman and Hall, London.

KENT, M. (1982). Plant growth problems in colliery spoil reclamation – a review. *Appl. Geog.*, **2**, 83–107.

LAX, A., DIAZ, E., CASTILLO, V. and ALBALADEJO, J. (1994). Reclamation of physical and chemical properties of a salinized soil by organic amendment. *Arid Soil Res. Rehab.*, **8**, 9–17.

MOFFATT, A. and MCNEILL, J. (1994). *Reclaiming disturbed land for forestry*. HMSO, London.

PALMER, J.P. (1984). *An investigation of the potential for the use of legumes on colliery spoil*. PhD thesis, University of York, York.

RICHARDS, I.G., PALMER, J.P. and BARRATT, P.A. (1993). *The reclamation of former coal mines and steelworks*. Elsevier, Amsterdam.

ROGERS, S.L. and BURNS, R.G. (1994). Changes in aggregate stability, nutrient status, indigenous microbial populations, and seedling emergence, following inoculation of soil with *Nostoc muscorum*. *Biol. Ferility Soils*, **18**, 209–215.

SCULLION, J. (1994). *Restoring farmland after coal: the Bryngwyn Project*. British Coal Opencast Executive, Mansfield.

SHILDRICK, J.P. (1984). *Turfgrass manual*. National Turfgrass Council, Bingley.

SIMMONDS, S.J., MAJER, J.D. and NICHOLS, O.G. (1994). A comparative study of spider (Araneae) communities of rehabilitated bauxite mines and surrounding forest in the southwest of Western Australia. *Restoration Ecol.*, **2**(4), 247–260.

STEVENSON, F.J. (1982). Organic forms of nitrogen. In Stevenson, F.J. (ed.) *Nitrogen in agricultural soils*. American Society of Agronomy, Wisconsin.

STEWART, V.I. and SCULLION, J. (1989). Principles of managing man-made soils. *Soil Use Man.*, **5**(3), 109–116.

URESK, D.W. and YAMAMOTO, T. (1986). Growth of forbs, shrubs, and trees on bentonite mine spoil under greenhouse conditions. *J. Range Man.*, **39**(2), 113–117.

WALTON, B.T., GUTHRIE, E.A. and HOYLMAN, A.M. (1994). Toxicant degradation in the rhizosphere. In Anderson, T.A. and Coats, J.R. (eds) *Bioremediation through rhizosphere technology*, pp. 11–27. Proceedings from a symposium in Chicago, Illinois, August 23–27, 1993. American Chemical Society Series 563, Washington DC.

WELSH DEVELOPMENT AGENCY (1993). *Working with nature*. WDA, Cardiff.

WILLIAMS, P.J. (1975). Investigations into the nitrogen cycle in colliery spoil. In Chadwick, M.J. and Goodman, G.T. (eds) *The ecology of resource degradation and renewal*, pp. 259–274. Blackwell, Oxford.

YOUNGER, A. (1989). Factors affecting the cropping potential of reinstated soils. *Soil Use Man.*, **5**(4), 150–154.

Further reading

HESTER, R.E. and HARRISON, R.M. (1994). *Mining and its environmental impact*, Issues in Environmental Science and Technology, No. 1. Royal Society of Chemistry, Cambridge.

WILLIAMSON, N.A., JOHNSON, M.S. and BRADSHAW, A.D. (1982). *Mine wastes reclamation*. Mining Journal Books, London.

There is also a series of reports and guidance documents from the Minerals and Land Reclamation Division of the UK Department of the Environment, published by HMSO, London:

Cost effective management of reclaimed derelict sites (1990)

Landform replication as a technique for reclamation of limestone quarries (1992)

The potential of woodland establishment on landfill sites (1993)

Landscaping and revegetation of china clay wastes (1993)

The reclamation and management of metalliferous mining sites (1994)

Restoration of damaged peatlands (1995)

Slate waste tips and workings in Britain (1995)

The journal *Land Contamination and Reclamation* (EPP Publications) has many review and research articles pertinent to this field.

Chapter 6

The establishment, management and maintenance of self-sustaining vegetation

6.1 Introduction

We may assume for what is about to follow that the land has had any un-
naturally arising contamination removed or transformed but we do not exclude
substrates that have a natural tendency to produce more chemical or physical
problems as they age. Essentially, we will be considering 'non-aggressive' materials as a starting point, for the purposes of establishing amenity reclamation or
a self-sustaining restoration.

The material may have arisen from three major categories:

1. Agro-forestry (e.g. arable crop land).
2. Short-term civil engineering disturbances (e.g. open-cast coal-mining).
3. Long-term industrial uses (e.g. abandoned gasworks).

Once the starting material has been treated it will be fit for starting a programme of plant establishment and management.

6.1.1 Land-use selection criteria

A decision as to what type of habitat is going to be recreated has to be made.
The decision will invariably involve a political element since the local authority
will be wholly or partially responsible for the implementation of the scheme.

Essentially the choice of use is between amenity or conservation. It is important to note here that the term conservation is not being used in its strictest
sense as the habitat has been lost and is now being restored, but is a convenient
term for public consumption.

Most reclamation and all restoration schemes involve the establishment of
vegetation. The vegetation may range from small planting beds on light industrial parks through the establishment of gardens to agricultural reclamation (cf.
Chapter 5) or the establishment of country parks or wildlife reserves. In many
instances the success of vegetation establishment is crucial to the public percep-

tion of the success of the reclamation scheme as a whole even if the major works have been in soil decontamination prior to vegetation establishment. Equally, vegetation performs a crucial engineering function at many sites (Fig. 6.1). Even if wildlife habitat creation is not the primary function of a reclamation scheme, the establishment of any vegetation provides an opportunity for the creation and enhancement of habitats of wildlife value.

The objective of the establishment of vegetation in a reclamation scheme should be for the vegetation to perform a function within the scheme. The function of the vegetation will be defined by the use to which the site is to be put and the characteristics of the site and its substrates. The type of vegetation established, the use to which the site is to be put and the characteristics of the site will all determine the nature of the establishment, monitoring and management practices that have to be adopted. During reclamation and restoration scheme design, ignoring the link between site characteristics, site use, vegetation type, monitoring and management is likely to lead to vegetation under-performance or even failure.

In this chapter the requirements of vegetation will be discussed in the context of the substrate types found on sites to be reclaimed or restored with the emphasis on restoration to end-uses other than agro-forestry.

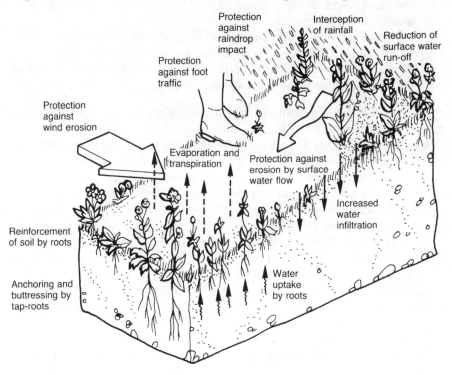

Fig. 6.1 Some influences of vegetation on the soil (source, Coppin and Richards, 1990).

6.2   Natural vegetation of degraded land

6.2.1   Succession

The pioneers of successional theory in the United Kingdom seemed quite clear that succession was the process of colonization followed by a series of stages where habitat interactions became more complex, leading to a climax community restricted from progressing any further by climatic or edaphic factors. The cycling at climax community stage was by a process of secondary succession following a cycle-initiating event such as the death of an oak tree in a climax oak woodland. Succession could be maintained at one of its seral stages by intervention by man, for example, by the introduction of grazing or mowing of grassland. Later thinking led to other models of succession which did not follow the classical model (Connell and Slatyer, 1977). These authors suggest that there are four modes of succession which may operate on a particular site:

1. **Facilitation** where early pioneer plants alter soil and site characteristics such as to make it suitable for other species to establish, i.e. those that are less tolerant to the initial harsh conditions.
2. **Tolerance** where slower-growing competitors eventually exclude the early dominant ruderal species.
3. **Inhibition** where pioneers exclude other plants until they become extirpated naturally, allowing the establishment of competitive species.
4. **Random**, the chance survival of different species on the same site which *de facto* has no adverse chemical status.

How the first three mechanisms may operate when a gap is opened up by a disturbance is shown in Fig. 6.2. We may regard the newly reinstated substrate of a disturbed substrate as such a gap, making this theory not only applicable, but eminently *testable* on degraded and disturbed sites. Furthermore, this theory is not inconsistent with the model proposed by Grime (1979); indeed, Grime's theory goes a long way to identifying the mechanism by which it may operate. Application of any of these models to derelict land has not been easy as some derelict land substrates seem to exhibit classical succession patterns and others do not. Moreover the assemblages of plants and animals found on derelict land substrates do not necessarily correspond with those found on natural substrates. So, for example, on colliery spoil although Hall (1957) suggested a clear successional series from observations of spoil heaps in the United Kingdom (Fig. 6.3), other workers looking at species number have found no correlation between species number and time since tipping but have found a correlation between species number and pH (Fig. 6.4). Petit (1982) working in France suggested a model which included spoil pH as a parameter (Fig. 6.5). Interestingly Petit suggests that colonization by trees particularly birches (*Betula* spp.) can occur as a first stage, a phenomenon occasionally seen on colliery spoil heaps in the United Kingdom.

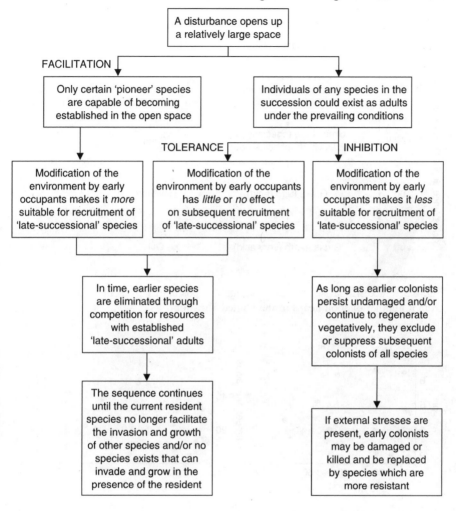

Fig. 6.2 Connell and Slatyer's (1977) model of mechanisms underlying succession.

Observations, again on colliery spoil, have led to a clearer understanding of the relationship between species composition of the vegetation on a site and that of the seed rain and seed bank (Huby, 1981). Older sites were found to have no greater number of species in the seed bank, seed rain or vegetation but these parameters did increase with increasing pH. Significant correlations were found between seed rain and seed bank numbers for individual species and all were affected by pH (Fig. 6.6). It was clear also that pH was not the only factor affecting seed-bank composition, and that toxicities associated with low pH and the ameliorating effect of factors such as the presence of organic matter vegetation cover modified simple pH effects (Table 6.1).

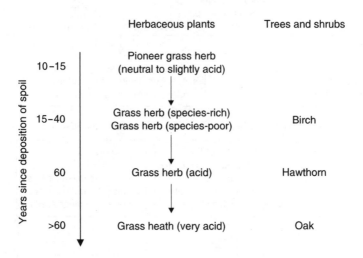

Fig. 6.3 Successional sequences of vegetation postulated by Hall (1957) using data from a number of spoil heaps in the United Kingdom.

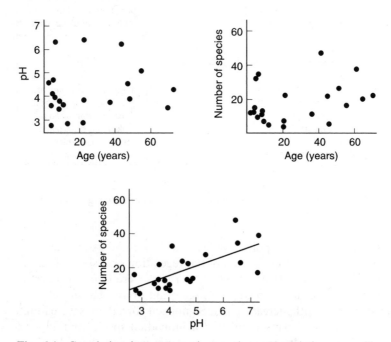

Fig. 6.4 Correlation between species number, pH and time on colliery spoil (after Bradshaw and Chadwick, 1980).

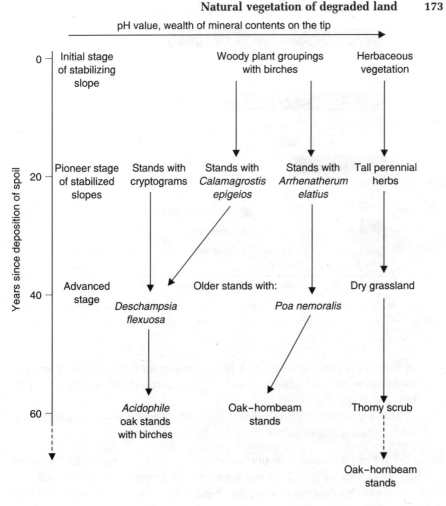

Fig. 6.5 Suggested successional sequences on the spoil heaps of Nord-Pas de Calais (after Petit, 1982).

The process of succession is essentially a process of overcoming limiting factors. This can be achieved by natural succession (a strong candidate for the 'facilitation' model of Connell and Slatyer, 1977), or by artificially accelerating the process by intervening to achieve the facilitation by artificial means. In Table 6.2 such factors are summarized from both the habitat and species perspectives, linking them to derelict land issues. It is perhaps unsurprising, considering the issues in Table 6.2 (and see Figs 6.7 and 6.8), that even on one substrate, colliery spoil, there should be the different findings expressed above. Both time and pH govern the process of succession on colliery spoil as follows:

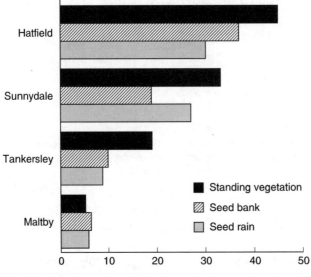

Fig. 6.6 Numbers of species in the seed rain, seed bank and standing vegetation at four sites in the Yorkshire coalfields (Huby, 1981). See Table 6.1 for site pH.

1. Where pH is low there will be little succession until the effects of low pH are ameliorated and colonization will be by those plants that can tolerate low pH only.
2. Where pH is not limiting colonization will follow a successional pattern with species composition dominated by local species.

Huby (1981) in fact concluded that succession on colliery spoil fitted Connell and Slatyer's model (Fig. 6.2) as she found no occurrence of species in the seed bank or rain of the sites she studied that were not able to survive on the sites as adults.

There has been much study of the natural colonization of china clay waste in relation to nutrient accumulation leading to suggestions of a capital of nitrogen

Table 6.1 The relation between pH, seed bank and vegetation cover at four sites in the Yorkshire Coalfield (Huby, 1981)

	pH	Viable seed as % of seed rain	% cover	% of annual germination of viable seed bank
Tankersley	4.2 ±0.49	47	90	1.6
Sunnydale	5.35 ±1.22	22	70	5.9
Hatfield	5.6 ±1.33	10	45	20.7
Maltby	3.37 ±0.3	8	0.5	0

Table 6.2   Factors limiting succession (after Bradshaw, 1993)

Factor	Consideration for derelict land
Species level	
Getting there	Plants and animals must be able to arrive at a new area for succession to start. This has two components: being available in the vicinity and being able to disperse effectively. For large expanses of derelict land there will be barriers to dispersal over the whole area for many species and dispersal may be only to the edges first with gradual dispersal to the rest of the site. More significant perhaps is the issue of availability in the vicinity. If derelict land is in an industrial area where there are insignificant tracts of natural vegetation, primary colonists may be very limited in their diversity.
Establishing	Plants have to germinate. Quick germination is often postulated as best in colonizing situations but delayed germination until conditions are favourable may aid species in unfavourable conditions on derelict land. Germinated plants also need to be able to tolerate the conditions the site presents and has to be able to tolerate all the conditions of stress on the site. These may be many and varied but they all have to be tolerated or the colonizing species will be selected out. For example plants colonizing metal mines have to tolerate high concentrations of metals *and* nutrient deficiency to survive. Tolerance of just one factor will lead to death.
Growth	Water and nutrients are needed for growth. However limiting these are the plant has to acquire them in enough quantities to grow. Succession may be limited by just one nutrient and of the major nutrients nitrogen is often a primary limiting factor on derelict land. Phosphate is also important as nitrogen-fixing legumes will not be able to fix nitrogen without an adequate phosphate supply. It is not clear how water supply affects succession, but on derelict land the combination of a saline colliery spoil and low rainfall has been shown to restrict nitrogen fixation by sown legumes and the resultant build up of nitrogen capital and ingress of naturally colonizing species (Figs 6.7 and 6.8).
Habitat	
Initial characteristics	Poor physical and chemical characteristics may bias early stages of succession considerably. These are important considerations on derelict land. These characteristics of habitat will change as succession proceeds but bias through toxicity may continually affect habitat change.
Allogenic factors	Allogenic factors, i.e. those arising from outside the developing community of organisms, may have a profound effect. On derelict land, accumulation of soil-forming materials, nutrient accumulation, inputs of nutrients from the air and leaching away of toxic substances may have a profound effect on the limiting factors on succession.
Autogenic factors	Autogenic factors, i.e. those arising from the activities of the living organisms, have a profound effect on succession. Nutrient accumulation and cycling causing development of soil structure and resultant habitat change are essential for succession to take place on derelict land.
Progression	Habitats have to change for succession to proceed. Factors limiting succession may change with progression and assumptions must be carefully made with respect to which factors are affecting succession at any one time. On derelict land materials where a number of factors may be potentially limiting, understanding of succession can only be achieved if the factors operating at any one time are carefully considered.

Table 6.2   (contd.)

Factor	Consideration for derelict land
Species interaction	At the ecosystem level the interactions between species may be important. Interactions may be facilitative or inhibitory. An example of a facilitative effect on derelict land is the role of legumes in building up nitrogen levels allowing other non-nitrogen fixing species to thrive.

having to be accumulated before an ecosystem can 'function' (Marrs and Bradshaw, 1993). There are parallels in natural and man-made systems for this concept (Table 6.3). Although the amounts or rates of nitrogen to be accumulated will vary between substrates and conditions, the concept of a trigger of nutrient accumulation before an ecosystem really 'gets going' appears sound. The role of legumes in building up nitrogen capital during the early

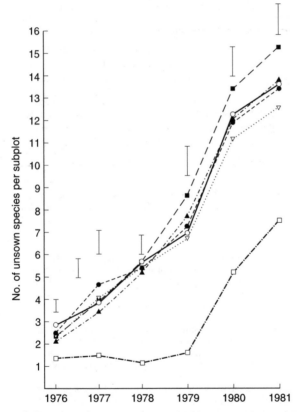

Fig. 6.7   Ingress of unsown species into legume and fertilizer plots on colliery spoil at Thorne, South Yorkshire. Symbols: □ legume plots, ▲, ▽, ■, ○, ●, various forms of applied nitrogen fertilizer. Least significant differences ($p = 0.05$) for each year are shown. See Fig. 6.8 for legume plot nitrogen levels.

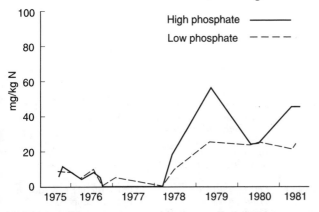

Fig. 6.8 Mineralizable nitrogen levels in legume plots at Thorne. Legume growth was inhibited by saline spoil 1975–1977 (source, Palmer, 1984).

stages of succession has been considered important on china clay waste and some natural substrates (Table 6.4). There is no doubt that if legumes are able to colonize and fix nitrogen, then nitrogen accumulation will be more rapid than otherwise; however, there is evidence that legumes or other nitrogen-fixing higher plants are not important colonizers of all derelict land substrates. Legumes are rarely colonizers of metal-mine wastes and are not important components of the primary succession on colliery spoils. Even though the presence of legumes has been widely documented on naturally vegetated colliery spoil (Table 6.5) there is little evidence that they have contributed significantly to nitrogen accumulation. This is probably due to the low phosphate status of colliery spoils and their ability to make unavailable phosphate that is

Table 6.3 Estimates of target nitrogen contents and the time taken to achieve these targets on four raw substrates (Marrs and Bradshaw, 1993)

Type of ecosystem	Time to develop non-nitrogen-fixing vegetation (years)	Nitrogen content (kg N ha^{-1})		Source
Glacial moraines	100	Soil (0–30 cm)	1200	Crocker and Major
		Litter	1000	(1955)
Sand dunes	21	Soil (0–10 cm)	400	Olsen (1958)
Ironstone spoils	100	Soil (0–21 cm)	600	Leisman (1957)
China clay waste	> 70	Soil (0–21 cm)	700	Roberts *et al.* (1981)
		Vegetation and litter	300	Marrs *et al.* (1983)
	> 120	Soil (0–21 cm)	1200	Roberts *et al.* (1981)
		Vegetation and litter	600	Marrs *et al.* (1983)

Table 6.4 Examples of nitrogen-fixing species which have been found on primary successions (Marrs and Bradshaw, 1993)

Substrate	Nitrogen-fixing species	Site location
Glacial moraines (Crocker and Major, 1955)	*Alnus crispa* *Dryas drummondii*	Glacier Bay, Alaska, USA
Glacial moraines (Viereck, 1966)	*Astragalus alpinus* *Astragalus tananaica* *Astragalus nutzotinesis* *Dryas drummondii* *Dryas integrifolia* *Sheperdia canadensis*	Muldrow Glacier Alaska, USA
Glacial moraines (Friedel, 1938a, b: Richard, 1968)	*Lotus corniculatus* *Trifolium badium* *Trifolium thalli*	Hintereisferner and Aletsch Glacier, Europe
Glacial moraines (A. D. Bradshaw, unpubl. data)	*Coriaria* spp.	New Zealand
China clay sand waste (Roberts *et al.*, 1981)	*Lotus corniculatus* *Lupinus arboreus* *Sarothamnus scoparius* *Ulex europaeus* *Ulex gallii*	Cornwall, UK
Ironstone spoils (Leisman, 1957)	*Melilotus alba* *Trifolium repens*	Minnesota, USA

input. Nitrogen fixation by legumes on colliery spoil has been shown to be sensitive to phosphate supply (Fig. 6.9).

Nitrogen is of course not the only consideration in the early stages of succession and Table 6.2 points to the number and complexity of factors governing early succession and that the importance of individual factors will change with time. In a study of an area affected by a non-anthropogenic disturbance, the Mt St Helens volcanic eruption of 1980, Halvorson *et al.* (1991) reported that two common perennial lupine species, *Lupinus lepidus* and *L. latifolius*, which had colonized naturally, had a significant impact on soil carbon, nitrogen and microbial activity, primarily within the rhizosphere. The authors also suggested that as the establishment of individuals was patchy, this actually contributed to a heterogeneous and therefore diverse substrate, which could lead to significant biodiversity at later stages of succession. This effect is an important one, and the importance of establishing variability in substrate characteristics in restoration schemes cannot be over-emphasized. Ireland *et al.* (1994) demonstrated that the amount of woody and rock cover was positively correlated with small mammal, lizard and amphibian species richness and abundance on a coal-mine restoration in northwestern New Mexico.

All stages of succession will involve lower plants, fungi and bacteria and these groups of organisms may play an important role in the facilitation of

Table 6.5 Some records of naturally occurring legumes on colliery spoil (after Palmer, 1984)

Species	Reference	Location
Anthyllis vulneraria	Hall (1957)	
	Molyneux (1963)	South Lancashire
	Chadwick and Hardiman (1976)	Yorkshire
Cytisus scoparius	Chadwick and Hardiman (1976)	Yorkshire
Lathyrus montanus	Chadwick and Hardiman (1976)	Yorkshire
	Chadwick *et al.* (1978)	South Wales
Lathyrus pratensis	Hall (1957)	
	Richardson *et al.* (1971)	Durham
	Chadwick *et al.* (1978)	South Yorkshire
	Huby (1981)	South Yorkshire
Lathyrus sylvestris	Brierley (1956)	Nottinghamshire, Derbyshire, South Yorkshire
Lespedeza thunbergii	Schramm (1966)	Pennsylvania
Lotus corniculatus	Brierley (1956)	Nottinghamshire, Derbyshire, South Yorkshire
	Hall (1957)	
	Molyneux (1963)	South Lancashire
	Richardson *et al.* (1971)	Durham
	Down (1973)	Somerset
	Chadwick and Hardiman (1976)	Yorkshire
	Gwent County Council (1976)	Gwent
	Chadwick *et al.* (1982)	South Wales, South Yorkshire
	Huby (1982)	South Yorkshire
Lupinus sp.	Chadwick and Hardiman (1976)	Yorkshire
Medicago lupulina	Brierley (1956)	Nottinghamshire, Derbyshire, South Yorkshire
	Hall (1957)	
	Chadwick and Hardiman (1976)	Yorkshire
	Gwent County Council (1976)	Gwent
	Elias *et al.* (1982)	Northumberland, Nottinghamshire
	Huby (1981)	South Yorkshire
Melilotus alba	McDougall (1918)	Illinois
	Croxton (1928)	Illinois
	Schramm (1966)	Pennsylvania
Melilotus officinalis	Hall (1957)	
Robinia pseudacacia	Schramm (1966)	Pennsylvania
Trifolium arvense	Hall (1957)	
Trifolium campestre	Hall (1957)	
Trifolium dubium	Hall (1957)	Yorkshire
	Chadwick and Hardiman (1976)	South Wales
	Elias *et al.* (1982)	
Trifolium hybridum	Chadwick *et al.* (1978)	South Wales, South Yorkshire
Trifolium pratense	Brierley (1956)	Nottinghamshire, Derbyshire, South Yorkshire
	Hall (1957)	
	Molyneux (1963)	South Lancashire
	Richardson *et al.* (1971)	Durham
	Down (1973)	Somerset
	Chadwick and Hardiman (1976)	Yorkshire
	Elias *et al.* (1982)	Northumberland, West Yorkshire, Nottinghamshire
	Huby (1981)	South Yorkshire

Table 6.5 (*cont'd*)

Species	Reference	Location
Trifolium repens	Brierley (1956)	Nottinghamshire, Derbyshire, South Yorkshire
	Hall (1957)	
	Molyneux (1963)	South Lancashire
	Schramm (1966)	Pennsylvania
	Richardson *et al.* (1971)	Durham
	Down (1973)	Somerset
	Chadwick and Hardiman (1976)	Yorkshire
	Gwent County Council (1976)	Gwent
	Chadwick *et al.* (1978)	South Yorkshire
	Elias *et al.* (1982)	Northumberland, West Yorkshire, South Wales, South Yorkshire
	Huby (1981)	South Yorkshire
Ulex europaeus	Richardson *et al.* (1971)	Durham
	Chadwick and Hardiman (1976)	Yorkshire
	Gwent County Council (1976)	Gwent
Vicia sativa	Chadwick and Hardiman (1976)	Yorkshire

colonization by higher plants. On mineral substrates bacteria, lichens and bryophytes may be important in weathering and exfoliation of rock, so facilitating build-up of fine-grained particles in which other plants can establish. Bartuska and Lang (1981) investigated factors affecting the rate of leaf litter decomposition in two areas, one a surface mine reclaimed in the 1940s, the other a mixed stand of maple and sycamore which had been partially cleared of maples at the same time. They found that the composition of the litter – not the numbers of decomposers or the microclimate – was important in controlling the rate of decomposition, in other words the effects of mining over clear cutting had by now disappeared.

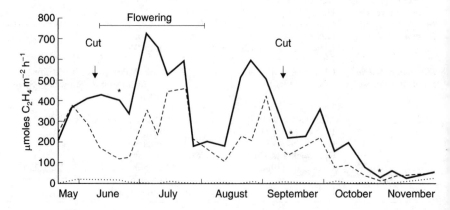

Fig. 6.9 Nitrogen fixation by white clover on reclaimed colliery spoil at two phosphate levels at Thorne, South Yorkshire (after Palmer, 1984): - - - low phosphate; —— high phosphate; * significant difference at $p < 0.05$.

Table 6.6 Examples of former industrial sites in Britain which have acquired conservation value status on botanical ground (from Johnson, 1978)

Location	Origin	Distinguished botanical features
Brickfields, Buckinghamshire[a]; Sanford, Berkshire[a]	Brick-clay extraction Sand extraction	*Blackstonia perfoliata, Lotus tenuis, Ophrys apifera, Dactylorhiza incarnata, Equisetum variegatum, Epipactis palustris, Potamogeton coloratus*
Aston Clinton, Buckinghamshire[a]	Chalk ragstone quarrying	*Hippocrepis comosa, Ophioglossum vulgatum, Polygala calcurea*, various Orchidaceae
Hurley, Berkshire[a] Millers Dale, Derbyshire[b,c]	Chalk quarrying Limestone quarrying/calcining	*Anacamptis pyramidalis, Gymnadenia conopsea, Ophrys apifera, Botrychium lunaria, Daphne mezerem, Parnassia palustris, Saxifraga tridactylites*, various Orchidaceae
Wingate, Durham[b]	Magnesian limestone quarrying	*Botrychium lunaria, Crepis mollis, Gentinella amarella, Saliz nigricans*, various· Orchidaceae
Honister Crag, Cumbria[d]	Slate quarrying	*Alchemilla alpina, Saxifraga aizoides, Oxyria digyna, Sedum rosea*

[a] Kelcey (1975).
[b] Davis (1976).
[c] Bradshaw (1977).
[d] Ratcliffe (1974).

It is not only plants and soil animals which may have beneficial effects with regard to re-establishment on nutrient cycling on reclaimed areas. Dixon and Hambler (1993) reported on the effects of rabbits on a limestone quarry bank, near Skipton, Yorkshire. The authors reported that there was increased nutrient content, moisture and organic material in middens (persistent dung-heaps) as compared to other areas of the site. This contributed locally to improvement of performance of pioneer plant species on the middens, the plants being greener, and having higher concentrations of nitrogen, phosphorus and potassium than in the surrounding area. This would also bring an increase in patchiness of the site, and the rabbits were judged to be 'beneficial' to the reclamation. As the objective of the scheme was to encourage a grassland community as the restoration end-point, and not trees, rabbits could probably be allowed to remain on the site.

6.2.2 Vegetation types

Some indication of the types of vegetation found on naturally colonized, former degraded sites is given in section 6.2.1. Substrate characteristics are principal determinants of the type of vegetation found on naturally recolonized degraded sites and can result in the presence of unusual vegetation types and

Table 6.7 Review summary of metallophyte vegetation in England and Wales (after Department of the Environment, 1994)

Species	Type	Habitat	Rarity/importance	Source reference
Thlaspi caerulescens (alpine pennycress)	Absolute metallophyte	Lead/zinc mine spoil, England (Pennines, Mendips) and N. Wales. Very rare in Scotland as arctic-alpine relic	Local and rare; very disjunct distribution	Ingrouille and Smirnoff (1986); Baker and Proctor (1990)
Minuartia verna (spring sandwort)	Local metallophyte	Lead/zinc (and copper) mine spoils on Carboniferous limestone. Also uncontaminated sub-alpine grassland	Locally abundant on suitable disturbed mining ground	Baker and Proctor (1990)
Silene vulgaris ssp. maritima (sea campion)	Local metallophyte	Lead/zinc (and copper) mine wastes (Mendips, N. Wales) and contaminated river gravels (central Wales and Northumberland). Primarily coastal; also arctic-alpine	Locally abundant but very limited inland distribution	Baker and Dalby (1980); Baker and Proctor (1990)
Armeria maritima (thrift)	Local metallophyte	Lead/zinc (N. Pennines) and copper (Cornwall, central Wales) mine wastes and contaminated river gravels. Primary distribution as for sea campion	Locally frequent at N. Pennine sites but rare in Wales	Baker and Proctor (1990)
Cochlearia pyrenaica (scurvy grass)	Local metallophyte	Lead/zinc mine spoil and river gravels in the Pennine orefield. Also montane (arctic-alpine) distribution	Very restricted distribution, usually close to water courses	Baker and Proctor (1990)
Viola lutea (mountain pansy)	Local metallophyte	Closed turf over lead/zinc mine spoils on Carboniferous limestone (Pennine orefield)	Often abundant. Various flower colour forms; largely yellow (S. Pennines) purple (N. Pennines)	Baker and Proctor (1990)
Dianthus deltoides (maiden pink)	Local metallophyte	Closed turf over calcareous lead/zinc mine spoil in Peak District	Rare. Uncommon elsewhere on dry sandy soils in lowland Britain	Baker and Proctor (1990)
Epipactis leptochila (slender-lipped helleborine)	Local metallophyte	Metalliferous gravels on R. South Tyne. Elsewhere on calcareous soils in S. England	Rare	Richards and Swan (1976)

Table 6.7 (contd.)

Species	Type	Habitat	Rarity/importance	Source reference
Epipactis youngiana (young's helleborine)	Absolute metallophyte	Lead/zinc mine waste and gravels, R. South Tyne	Very rare; only two sites recorded	Richards and Porter (1982)
Epipactis helleborine (broad-leaved helleborine)	Local metallophyte	Lead/zinc mine spoils, S. & N. Pennines	Widely distributed throughout British Isles, but rare on metalliferous spoil	Holliday and Johnson (1979)
Epipactis atrorubens (dark-red helleborine)	Local metallophyte	Lead/zinc mine spoil, S. Pennines	Rare. Primarily a species of limestone cliffs and pavements in Pennines and N. Wales	Holliday and Johnson (1979)
Asplenium septentrionale (forked spleenwort)	Local metallophyte	Siliceous rocks, often associated with derelict lead mine workings in N. and central Wales	Uncommon	Baker and Proctor (1990)
Botrychium lunaria (moonwort)	Local metallophyte	Closed turf over lead/zinc mine spoil in Pennines. More typically a species of open dry grasslands	Frequent but often overlooked	Baker and Proctor (1990)

Table 6.8 Species found on over 25% of the colliery spoil sites in a survey of 15 sites in England and Wales in 1976 (source, Chadwick *et al.*, 1978)

Species	% of sites
Common bent (*Agrostis capillaris*)	100
Wavy hair grass (*Deschampsia flexuosa*)	53
Yorkshire fog (*Holcus lanatus*)	53
Red fescue (*Festuca rubra*)	40
False oat grass (*Arrhenatherum elatius*)	27
Cocksfoot (*Dactylis glomerata*)	27
Blackberry (*Rubus fruticosus* agg)	53
Gorse (*Ulex europaeus*)	27
Rosebay willowherb (*Chamerion angustifolium*)	60
Sheep's sorrel (*Rumex acetosella*)	60
Hawkweeds (except mouse-ear)	47
Mouse-ear hawkweed (*Hieracium pilosella*)	40
Cat's ear (*Hypochaeris radicata*)	40
Coltsfoot (*Tussilago farfara*)	33
Birdsfoot trefoil (*Lotus corniculatus*)	27
Birch (*Betula pendula*)	27
Oak (*Quercus petraea*)	27

rare species (Tables 6.6 and 6.7). Substrates such as colliery spoil do not generally support nationally rare or uncommon vegetation types but may in industrialized areas represent one of the few areas of semi-natural vegetation, and may represent or support species which are no longer found in the vicinity. Table 6.8 indicates the occurrence of the most common species on colliery spoil in England and Wales, and Table 6.9 the types of vegetation associated with colliery habitats. Even though such sites do not generally support vegetation of great interest old colliery spoil heaps in Yorkshire are vegetated with acid oak woodland and upland spoil heaps in South Wales with acid heath and heathy grassland. Interestingly the Tankersley Lidgett site studied by Huby (Table 6.1) supports oak woodland with a wavy hair grass understorey and is an example of an old site with a very acid spoil but supporting a recognized vegetation type. Although species numbers are low this feature is typical of this vegetation type and hence a complicating factor in considering pH and species number relationships in relation to succession.

The increase in the amount of derelict land, left undeveloped for some years, and being colonized by plants and animals has led to the study of the resulting plants and animals and the introduction of the concept of the 'urban common' as a distinct community of plants and animals different from that found in rural areas. Gilbert (1989) has described a succession on such commons based mainly on work in South Yorkshire (Table 6.10). Although this work was based mainly on demolition sites where demolition rubble predominates it provides useful pointers to the processes taking place. In particular Gilbert considered the heterogeneity of the site and the role this plays in vegetation development in relation to Grime's model of vegetation succession (Grime,

Table 6.9 Habitats and vegetation recorded in surveys of all derelict land in Fife in 1980 and 1988 (source, Hellings, 1988)

Vegetation	Substrate type	Comments
Bare ground	Concrete	Cracks developing being colonized by ruderal plants
	Slurry beds	Lack of vegetation confined to areas which become waterlogged – indicates that waterlogging is prime cause of vegetation absence
	Steep colliery spoil slopes	Constant movement of surface material and high surface temperatures principal cause of lack of vegetation
Open herb communities	Building rubble	Rapid colonization of these substrates, at one colliery developing into grassland in eight years. The calcareous and diverse nature of these materials and lack of toxicity often lead to species of some wildlife value establishing
	Railway clinker	50% ground cover in five years beginning to be followed by birch and willow
	Slurry beds	Only colonized on the drier areas. Very few species in early colonization
	Steep colliery spoil	Open herb and scrub vegetation common pioneers. Little evidence of colonization by species of particular note
	Flat colliery spoil	Although providing more favourable conditions for vegetation establishment than sloping sites vegetation establishment is still slow. Stands of some less common species (e.g. teasel, *Dipsacus fullonum*) are found and at one site surrounded by a semi-natural meadow a wide range of species indicating the influence of nearby sources of seed on species composition
Bare ground colonized by scrub	Railway clinker	Birch easily colonizing this material
	Slurry beds	Birch pioneer vegetation at one site has colonized 2 ha of land between 1980 and 1988. Area surrounded by dense woodland
	Colliery spoil	Birch is the most common pioneer but elder, ash, willow and rowan were all found colonizing spoil materials usually at the base and lower sides of tips. At two sites with long established scrub cover on the sides Scots pine, willow, sycamore, birch and even oak were beginning to colonize the tops of the tips

Table 6.9   (contd.)

Vegetation	Substrate type	Comments
Grassland and tall herb communities	All substrates	One or two older sites had developed meadow vegetation and had been colonized by some less common species such as orchids and yellow rattle. Grazing was considered to be important to maintain these communities
Grassland and open communities colonized by scrub	All substrates	Once herbaceous vegetation had developed, scrub was observed to colonize at a number of sites. Species included the trees mentioned above and broom, gorse and hawthorn

Table 6.10   Succession on urban commons (after Gilbert, 1989)

Successional stage	Description
Oxford ragwort	Short-lived perennial plant stage dominated by Oxford ragwort and other species with windborne seeds. Colonization is aided by the temporary use of such sites as car parks with Black Twitch grass apparently introduced preferentially to such sites.
Tall herb	After 3–6 years vegetation becomes dominated by tall perennial with leafy stems and proportion of annual rosette plants falls. Rosebay willowherb is a characteristic species of this stage. The stage is species rich but variable in composition both between and within sites. Calcifuges are, however, rare.
Grassland	After 8–10 years perennial grasses begin to dominate with only relatively small patches of tall herbs left. Many of the remaining herbs spread by vegetative means. Japanese knotweed may also become dominant at this stage. At this stage also bryophytes, fungi and lichens may be abundant.
Scrub woodland	Colonists such as birch or willow with windblown seeds can enter the succession early on, in fact for these species entry early is more likely than later unless there is disturbance of the site. However, larger seeded trees and shrubs such as ash, sycamore and hawthorn enter later and have the capacity to enter closed vegetation communities. They are aided by dispersal by animals and birds attracted to the site by existing vegetation. The resulting woodland is often unlike other self-sown woodlands, being a combination of native species such as ash, hawthorn and willow and introduced or cultivated species such as laburnum, domestic apple and garden privet.

Table 6.11   Regional factors influencing the nature of urban commons (after Gilbert, 1989)

City	Urban common characteristics
Bristol	Large amounts of buddleia which is probably succeeded by sycamore but which can dominate sites for many years. Traveller's joy is also common. There is also a maritime influence so species such as sea pea are often found on appropriate substrates in the city. Rosebay willowherb not as common as in many other cities.
Swansea	Large amounts of Japanese knotweed. Buddleia and hemp agrimony also common. Like Bristol, influenced by its maritime position and rosebay willowherb not common.
Liverpool	Woody plant colonizers rarely seen and many sites are in early stages of succession because of disturbance.
Manchester	Substantial amounts of Japanese knotweed on most commons. Wetland species such as reed canary grass also common.
Swindon	Species-rich neutral or calcareous grassland is typical of many sites in addition to railway species such as toadflax.
Hull	False oat grass and hogweed dominate many of the older sites in contrast to Sheffield where large umbellifers such as hogweed are not common.
Sheffield	Typified by a large number of garden escapes, such as michaelmas daisies and lupins which are very colourful. Eastern rocket, wormwood and Yorkshire fog also abundant. Goat willow is the principal woody colonizer.
Birmingham	Little buddleia but golden rod abundant. Larger bindweed also common.

1979). For example succession appears to follow different patterns on different materials:

1. Productive coarse rubble: ruderals → competitive ruderals/competitors → stress-tolerant competitors.
2. Rubble mixed with inert materials: ruderals → stress-tolerant ruderals → stress tolerators.
3. Crushed brick–mortar rubble: ruderals → C-S-R strategists (generalists) adapted to moderate levels of stress and disturbance.

Disturbance and chance introductions such as in garden refuse during succession will alter successional direction and change the end-point vegetation. The sensitivity of successional direction and end-point vegetation to site heterogeneity and disturbance that this work shows applies to most types of derelict land. Successional direction and vegetation type in urban commons are also influenced by regional factors which include not only factors such as climate but also the character of cities and their industries (Table 6.11).

6.3   Nature conservation

The characteristics of a site and the way it is managed are determining factors in the nature conservation value of a reclaimed site, irrespective of the use to which the site is put. There are three broad ways in which wildlife conservation needs can be introduced into a reclamation scheme:

1. Existing sites of nature conservation value can be retained within the scheme.
2. Reclamation schemes can incorporate features which provide a potential for wildlife value to develop.
3. Previously reclaimed sites can be managed or treated to increase their wildlife value.

6.3.1   Existing nature conservation value

Many sites which may have been derelict for some years have developed features of wildlife value. Restoration schemes should aim to retain all or some of these features because:

1. They may be important features on a local, regional or national scale of which the loss would be a consequential reduction of wildlife value, not just of the scheme but of a wider area.

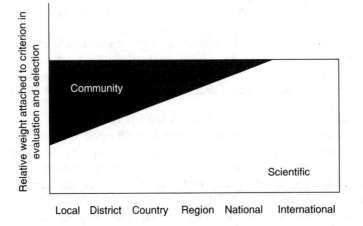

Fig. 6.10   The relationship between the importance of scientific (ecological) and community factors in the evaluation of sites of nature conservation importance (source, Collis and Tyldesley, 1993).

2. Retention of such features will provide a source of plants and animals for colonization of the remaining part of the reclamation scheme.
3. Retention of such features will provide continuity between the un-reclaimed, reclaimed and restored states of the site – this is often of importance where the wildlife status of the site has been valued by local people.

The last point is of relevance to the view that nature conservation is not only concerned with the ecological value of a site but also with community factors and that these community factors are of most importance at a local level (Fig. 6.10). The criteria for the evaluation of ecological value are well known and broadly agreed world-wide (Table 6.12). Although community factors such as amenity and education value are listed in Table 6.12 their use is less frequent than other factors. Some local authorities in the United Kingdom have included such factors in their evaluation frameworks for sites of local nature conservation interest (Table 6.13). There is, for example, much scope for the introduction of diverse woodland areas in urban landscapes (Hodge and Harmer, 1995).

There may also be geological features of importance and these should be treated in the same way as features of wildlife interest. Examples may be: interesting geomorphology, outcrops of unusual minerals or minerals with a historic connection (e.g. previously worked minerals in a former mining area) and secondary minerals developed in spoil heaps (e.g. internationally rare

Table 6.12 Criteria used in wildlife conservation evaluation in nine studies reviewed by Margules and Usher (1981), as adapted from Usher (1986)

Criterion	Frequency of use in nine published studies
Diversity (both of species and habitats)[a]	8
Naturalness[a]	7
Rarity (both of species and habitat)[a]	
Area (extent of habitat)[a]	6
Threat of human interference	
Amenity value	
Education value	
Recorded history[a]	3
Representativeness	
Scientific value	2
Typicality[a]	
Uniqueness	
Availability	
Ecological fragility[a]	
Management considerations	
Position in ecological/geographic unit[a]	1
Potential value[a]	
Replaceability	
Wildlife reservoir potential	

[a]Criterion used by the Nature Conservancy Council for assessing sites in the United Kingdom.

Table 6.13 Community value criteria taken into account by some UK authorities in their evaluation of sites of wildlife interest

Authority	Criteria
Greater Bristol	Community or amenity value, physical access, visual access, education value, aesthetic or landscape appeal, situation in area lacking natural habitats, recorded history
Derbyshire (outside National Park)	Amenity and education value
Greater London	Urban character, cultural/historic character, access, use, e.g. education, aesthetic appeal
West Midlands	Aesthetic appeal, educational/potential use, accessibility to public
Sheffield	Community value, educational value, isolation

secondary minerals formed since spoil deposition have been discovered at mine sites in mid-Wales).

6.3.2 Reclamation to increase wildlife value

Reclamation of a site has to satisfy a variety of needs and the design of a reclamation scheme often has to strike a balance between engineering constraints, landscape considerations, land-use needs and conservation needs. There is, however, almost always an opportunity to incorporate features of wildlife value or potential wildlife value and perform the necessary functions of the site. In addition to retaining existing sites of wildlife interest, features which will add wildlife value are as follows:

1. Incorporation of specific features such as water-bodies.
2. Treatment of features which are, say, a necessary part of the engineering function in a way which will enhance their wildlife potential, e.g. watercourses (Table 6.14; and Figs 6.11 and 6.12).
3. Provision of structural diversity to a revegetation scheme, e.g. use of hedgerows and tree planting, areas of different grassland management regimes, varying topography.
4. Use of substrate materials of differing characteristics, e.g. fertile/infertile, acid/alkali.
5. Establishment and management techniques aimed towards producing vegetation of a specific type which is of wildlife value (Table 6.15).

A particular difficulty in reclamation is that the requirement to produce a 'green' cover quickly conflicts with the aim of establishing vegetation of wildlife value especially where herbaceous swards are concerned. Hedgerows, copses,

Table 6.14 Measures to improve the ecological value of water courses (after National Rivers Authority, 1994)

Operation	Considerations
In-channel works	These include dredging, reprofiling and introduction of material to improve bed stability. Key considerations are: • keep excavation to a minimum except where system already degraded; • leave banks untouched wherever possible; • leave the channel as varied as possible; • work from one bank only to minimize disturbance, retain islands of vegetation to allow recolonization; • work from downstream up so colonization can occur on already worked on areas; • retain trees; • restore pools and riffles; • minimize disturbance to areas not being treated.
Riverbank works	These include resectioning and reprofiling. Where banks are of high wildlife interest alternatives should be considered. Where reprofiling takes place banks with a natural shape are preferred to straight angled batters. Banks should incorporate holes, ledges, crumbling slopes, shelves and shallow slopes (Fig. 6.11). Natural profiles should be restored, e.g. steep banks on outer edge of meanders, shallow slopes on inner edge.
Channel construction	These include construction of weirs, revetments, floodbanks, flood attenuation areas and channels. Major considerations include those for riverbank works above and the provision of diversity of habitat where possible. Figure 6.12 illustrates the influence of available land on habitat creation possibilities. In new channels the opportunity should be taken wherever possible to vary channel bed conditions to allow a range of bed conditions to develop and so diversify aquatic life. Variable bed conditions can be achieved by using different materials in different sections of the channel (e.g. materials of different particle sizes) and providing pools and faster flowing sections of water course.

water-bodies and water courses can be established which will increase in wild-life value; however, a monoculture of productive grasses required to stabilize the surface and provide a green cover will only very slowly move towards a more ecologically valuable sward. Sowing of wildflower mixes can be an alternative, but one that is more viable in the long term is to design the scheme so that the quick cover is provided where needed, but in other parts of a site a slower, but ecologically more valuable, approach is adopted. Tree planting can aid here by screening the more slowly developing parts of the site and by providing a ground layer which will colonize slowly as it will not have been sown to avoid competition between vigorous herbaceous plants and the trees.

Where a quick green cover is not essential a more natural approach to reclamation can be adopted (Welsh Development Agency, 1993). Here by a careful process of surveying and conserving naturally developed vegetation,

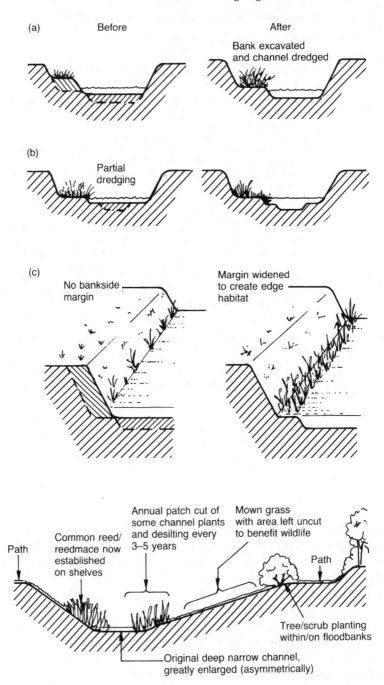

Fig. 6.11 Examples of some measures to enhance the wildlife value of banks of water courses (National Rivers Authority, 1994): (a) enhancing; (b) retaining; and (c) creating edge habitats (adapted from Brandon, 1989).

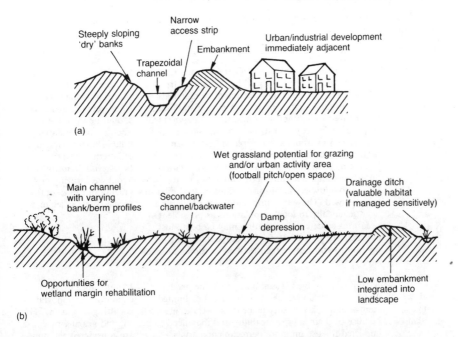

Fig. 6.12 The influence of available land on habitat creation possibilities (National Rivers Authority, 1994). (a) Narrow corridor. Access strip too narrow to use as amenity area. No potential for grazing; mowing is the only management option. (b) Wide corridor. Development pressure moved away from the river's edge. Viable size for amenity area. Potential for rehabilitation/re-creation of wetland margins, backwaters and wet grassland. Potential for grazing of wet grassland.

assessing the characteristics of the site and selecting plants which will grow satisfactorily with minimum of site amendment, a more natural vegetation cover can be established with limited expenditure. Such an approach would make use of on-site materials, readily available inexpensive amendments and natural colonization to achieve a diverse and robust vegetation which is also relatively inexpensive to produce.

6.3.3   Increasing the wildlife value of previously reclaimed sites

Previously reclaimed sites may have been sown to a monoculture of grasses with little consideration of the issues described in the previous section. These sites can be improved by the following means:

1.   Diversification by provision of ponds, hedgerows and trees.

Table 6.15 Examples of the requirements of the creation of habitats of wildlife value in reclamation schemes

Habitat type	Substrate and establishment requirements	Management requirements
Calcareous grassland	Requires alkaline substrate. Calcareous rock such as quarry waste can be improved by cultivating to 100 mm and introducing water-retentive materials such as organic matter or soil material. Establishment is improved by low fertilizer additions. Legumes will thrive on calcareous substrates and should be encouraged by application of phosphate. Open pioneer vegetation can be sown (e.g. a *Festuca* spp./ legume mix) and not prevent natural colonization. Should a suitable natural seed source not be available locally to allow colonization, wildflower seed mixtures may be used. These mixtures should be of plants as near genetically to local populations as possible. On occasions calcareous vegetation may be being lost elsewhere (e.g. as part of a limestone quarrying operation) and can be transplanted as turves to a new site.	Some 'weed' control may be necessary. The principal management tool is, however, grazing which should be introduced about two years after establishment. Grazing, although necessary, should be carefully controlled to avoid overgrazing and grazing out of desirable species such as legumes.
Wet meadow/ damp grassland	A high water table is essential and can be produced by impeded drainage, shallow gradients or limited compaction. Clayey, silty or peaty substrates are best because of their low permeability and may need to be imported for use on more permeable materials. Establishment may be by natural colonization if there are wet habitats nearby, otherwise commercial seed may be used. Where available, cut hay from wet meadows may be used as a seed source. Transplantation either by turves or disturbed soil and vegetation is possible but needs careful management and is only relevant where the donor site is to be damaged or lost.	Maintenance of wet conditions is important as is cutting or grazing. Hay cut and grazing by cattle are considered the most appropriate management regimes.
Heathland	An acidic, stable substrate is necessary. Although found naturally in very nutrient-deficient conditions principal heathland species such as *Calluna* and *Erica* spp. do benefit from seedbed nutrient additions. The surface must not be compacted for germination to occur and an uneven topography aids germination. Heathland establishment is slow and methods of establishment rely on collected heathland litter or harvested shoots which are then spread with a companion grass. Companion grasses used are generally *Festuca* and *Agrostis* spp. sown at a low rate. Turf and topsoil transfer can be achieved but again are only relevant where the donor site is to be lost.	Establishment is slow and sites should remain undisturbed by grazing or trampling during the establishment period which can be five years. Because of the infertile conditions few undesirable species become established but some trees particularly birch may colonize and should be removed.

Table 6.15 (*cont'd*)

Habitat type	Substrate and establishment requirements	Management requirements
Open water bodies	Calcareous substrates produce the most productive areas of water. The creation of shallows is a principal requirement for high productivity. Diversity of the lake bed in terms of materials and topography will also aid biological diversity. Any material can provide the form of the water-body but an impermeable layer will be needed to retain water. The impermeable layer is likely to be clay or even a membrane, so covering with coarser material will aid productivity. Topsoil or a layer of organic matter on the waterbody bed will aid rapid development of biological activity. Care must be taken not to encourage eutrophication. Bays, islands and peninsulas will all aid diversity and encourage birdlife. A range of emergent and aquatic vegetation may be established in the new waterbody.	Management is principally concerned with monitoring water quality, maintenance of bankside vegetation, removal of undesirable vegetation and managing site use.

2. Changes in management practices such as reduction of grazing at grazed sites or introduction of grazing at ungrazed sites.
3. Recultivation of moribund swards followed by resowing and adoption of management practices to encourage sward diversity.
4. Changes in site use, e.g. to reduce trampling pressure or compaction due to vehicular use.

Maunder (1992) has outlined the importance of reintroducing plants into sites from which they have become extirpated (removed completely) as part of a general strategy of biological conservation, and in this respect reclamation and restoration schemes have the potential for becoming nationally important. Gorsira and Risenhoover (1994) demonstrated, in a study of reclaimed mine sites near Fairfield, Texas, that it was difficult to establish native species in all situations, that it would require at least 27 years before planted specimens achieved the sizes of trees in undisturbed areas nearby, and that overall the survival rates were low. The authors suggested that without periodic intervention and replanting with variable species such as Washington hawthorn (*Crataegus phaenopyrum*) and water oak (*Quercus nigra*) in wet areas, and xeric species such as blackjack oak (*Q. marilandica*) on the dry upland, the success rate would remain low, and there would be little use made of the reclamation by mobile species, such as the white-tailed deer (*Odocoileus virginianus*). It is possible to obtain stock from adjacent undisturbed areas in some cases, as demonstrated by Bellairs and Bell (1993) in a restoration programme on a mine site in Western Australia. Ireland *et al.* (1994) suggested that the provision of rocks and vegetation patches would increase the diversity of small animals found on the site.

There are many examples of man-made sites in Central Europe which have been recolonized naturally by pioneer species. Prach and Pysek (1994) reported on the performance of woody plants on 15 successional series, starting with bare ground, in the Czech Republic and Hungary. The principal controlling factor as to plant establishment was retarded in dry, acid, nutrient-poor sites, and those species which reproduced principally by seed dispersal did better than others. They gave a number of recommendations for the establishment of plants on sites in the region, according to their disturbance history:

1. **Sites disturbed by mining**. Plant woody species on vulnerable slopes and margins, allowing the remainder of the site to recolonize naturally.
2. **Areas deforested by air pollution**. Because of problems with competition of rhizomatous species and the presence of pollutants, the removal of topsoil was suggested, then replanting with birch (*Betula pendula*), mountain ash (*Sorbus aucuparia*), and goat willow (*Salix caprea*). Aerial sowing was shown to be successful with birch.
3. **Urban areas**. Goat willow and European elder (*Sambucus nigra*) were found to be successful colonizers, which could then be enhanced for diversity by selective cutting and planting.
4. **Abandoned fields**. These sites were strong candidates for the establishment of species-rich pastures, as the herbs out-competed tree saplings effectively, although in the longer term trees could be established.

It is clear from this and many other studies that the birch is capable of being used as a pioneer species in a wide variety of situations.

6.4 Soil development

6.4.1 Naturally vegetated sites

Soil development on urban waste materials has been discussed. Soil development on other waste materials follows a similar process and the rate of development depends principally on those factors governing organic matter build-up and nutrient cycling. Where soil has developed in nutrient-poor conditions and particularly in upland areas, substantial peaty layers may be developed on the surface of the waste material with a layer of weathered material below but with relatively unweathered material near to the surface. The peaty layer is derived from dead vegetation which has accumulated over time, and whereas the material is of a wide carbon to nitrogen ratio and still nutrient-deficient, it is a much more hospitable material for root development than the underlying waste. Peaty material developed in such a way may also act as a sink for contaminants such as metals. For example, soils in North Wales affected by wind-blown pollution from metal mines have developed a peaty layer containing tens of thousands of parts per million of metals on which a closed vegetation cover including non-metal tolerant plants has developed.

There are also areas which have been 'abandoned' from agro-forestry, i.e. no longer managed for production, but not planted or cultivated for restoration either. This is fairly common in certain areas of the United States.

Mycorrhizal associations and the activities of soil organisms such as earthworms all assist in soil development as described in Chapter 2. Nelson and Allen (1993) noted that although the introduction of mycorrhizal symbionts was useful in establishing newly revegetated areas, where undesirable plants had already taken hold, normal weed control measures had to be put into effect.

6.4.2 Restored sites

There has been a considerable amount of research principally in association with the mineral extraction industry on the recovery or restoration of soil that has been removed during mineral extraction and replaced. Much of this work has been concerned with the return of the soil to a productive state for agriculture and the principles are dealt with in Chapter 5. One project, the Bryngwyn Project, involving experimenting with open-cast coal restoration techniques, was set up in 1977 and monitored for 17 years. The conclusions of the study included those relating to soil handling practices in relation to soil quality and are applicable to other sites (Table 6.16). Interestingly, conclusion 7 in Table 6.16, relating to the undesirability of over cultivation of the seed-bed, has also been concluded from restoration of soils after pipeline laying. Reducing soil particle size by cultivation after it has been stored and replaced greatly decreases the possibility of soil structure being quickly re-established and can lead to surface panning, particularly if heavy rain follows soil replacement and cultivation.

Where soils have not been replaced and are intended to develop from waste or spoil materials, research has centred on the effect of treatments on soil development. Again, more work has been done on colliery spoil than other materials and this work provides a guide to the factors important in soil development on mineral substrates (Table 6.17). Surveys of reclaimed colliery spoil sites have shown that soil development progressed much faster on some sites than on others (Department of the Environment, in preparation). Principal factors restricting soil development are:

1. Compaction and covers of soil which limit the depth to which weathering can occur.
2. Compaction and/or impeded drainage which restrict the incorporation of organic matter by root penetration or earthworms.
3. Spoil acidity reducing root penetration.

Table 6.16   Conclusions with respect to soil-handling from the Bryngwyn project (Scullion, 1994)

1. Topsoil storage eliminated earthworms from all but the surface of topsoil stores. The reduced volume of soil habitable by earthworms, a result of the anoxic conditions within such stores, ensures a marked reduction compared with populations present in undisturbed land.
2. Structural properties of soil from the inner core of topsoil heaps deteriorated during storage. Some damage was, however, associated with each phase of soil-handling operations.
3. Intensive cultivation following soil replacement had no persistent benefit in relieving soil compaction. It greatly decreased aggregate water-stability and earthworm abundance, and probably enhanced losses of nitrogen.
4. Recently replaced soils were found to be capable of supporting normal populations of earthworms, if suitably managed, and their activities led to a marked improvement in soil fertility. Over a six-year period, structural properties near the surface of soils with high earthworm densities improved rapidly, producing conditions close to those of undisturbed land. Sustained soil improvement depended, however, on the maintenance of an outlet for infiltrated water.
5. Where practicable, topsoil storage should be avoided. Where topsoil is stored, materials from the surface layer of these stores should be replaced in strips from which intervening areas can be colonized by earthworms.
6. Management of topsoil stores had a large effect on earthworm populations within those stores. High populations of beneficial, deep-burrowing species were favoured by frequent cutting and moderate nutrient inputs. Appropriate management of topsoil stores would greatly increase their potential for accelerating re-colonization of reinstated land, using the approach described in 5.
7. Cultivations following topsoil replacement should be restricted to the minimum required to produce a coarse seedbed and, if necessary, seeding rates should be increased to compensate for any reduction in percentage germination.
8. Minimum reinstatement gradients of 5% should be adopted for agricultural land in order to facilitate subsequent site drainage.

On other spoil materials there may be other factors which restrict root penetration in a similar way to acidity, for example high metal content or high salinity levels.

There was no evidence of spoil consolidation by weathering. Where high bulk density was found, it was attributed to compaction carried out as part of the initial placement of the spoil. Following such compaction, physical alteration of the compacted spoil did not occur.

Spoil cracking in compacted spoil permits root penetration. Spoil cracking is dependent on significant desiccation of the spoil which is more likely to occur where deep soil covers are not present, and where a substantial, actively transpiring vegetation cover is present.

Root penetration in compacted spoil follows pathways around stones or through stony material. Stony material in the upper spoil layers offers a benefit for drainage and aeration where excessive compaction may occur, since even at high bulk density the pore sizes present are sufficient to permit drainage. Lagoon fines are root-penetrable, but unless the rate of drainage can be improved, seasonal waterlogging is likely to kill vegetation. Soil development would be facilitated by cultivation if a stable soil structure could be maintained.

Table 6.17 Conclusions from studies of treatment of colliery spoil to improve its physical and biological properties (Rimmer and Colbourn, 1978; Colbourn and Stanton, 1979; Gildon *et al.*, 1982)

1. Soil physical conditions in colliery spoil are very poor compared with agricultural soils. Principal problems are caused by low porosity, resulting in poor water-holding capacity and impeded drainage.
2. The prospect of good soils developing on colliery spoil in the long term is not good. Comparison was made with the Dale soil series developed from Carboniferous shales which exhibit both waterlogging and low resistance to drought.
3. The results from relieving compaction by ripping were inconclusive. Ripping increased yields at one site but reduced them at another.
4. Bulky organic amendments can improve soil conditions.
5. Colonization of freshly exposed colliery spoil by earthworms was considered to be unlikely even in non-acidic conditions.
6. Soil amendments increased yields particularly on acidic spoils. Root penetration of underlying spoil was minimal on soiled sites.
7. Mixing of soil with spoil did not provide yield increases over a thin covering of the same soil although root penetration was deeper.

Where the development of a deep soil from colliery spoil is the long-term objective of reclamation these studies concluded that the following principles should be followed:

1. Place spoil without excessive compaction (bulk density ≤ 1.5).
2. Ensure root penetration is not impeded by restricted drainage or by acidity.
3. Stabilize spoil pores by incorporating humified organic matter and/or developing a vigorous, deep rooting system using a temporary crop if necessary.

6.5 Moving from agro-forestry to self-sustaining systems

When changing from agro-forestry to restoration or other reclamation uses there are some problems which have to be overcome before the desired species may be established:

1. High nutrient content of soils favouring ruderal species.
2. Low biological potential on site, with regards to plant-, animal- and soil-living species.
3. Possibly inappropriate drainage patterns and moisture-retaining characteristics.
4. Low substrate variability.
5. Lack of cover for animals.
6. Presence of artificial barriers to large animal movement.
7. High risk of erosion.
8. Lack of variability in stress and disturbance on-site.

All or just some of these will be features of any particular site and they must be assessed on a site basis, and will be addressed individually in the first instance.

6.5.1   Nutrient reduction

Nutrient-rich systems may be species-poor owing to dominance by a few, or even one, ruderal species. This is not always the case, for example when the soils are nutrient-impoverished and nothing will grow, but it does hold as a general rule. Consequently, some attempts have been made to reduce the nutrient contents of formerly fertile soils. This may be achieved by:

1. Allowing natural successional development.
2. Sowing native species capable of impoverishing the soil.
3. Continuous fallowing.
4. Physical treatment.

Crops may be continually taken without fertilization. This can run into trouble, however, under the law of diminishing returns, when the taking of the crop no longer is economical (i.e. pays for the harvester) when the management aim has not been reached. It is important under these circumstances to increase the incorporation of nutrients into slowly decomposing organic matter, and to encourage leaching (e.g. fallowing). There have also been several examples where biodiversity of vegetational assemblages has been improved by stripping off the top, nutrient-rich, soil layer. This leads to the establishment of a species-rich sward. Morgan (1994) described a programme where mixtures of sawdust and sugar were introduced into a prairie soil as a carbon source, resulting in the large-scale immobilization of nitrogen, and the suppression of weed species, allowing the native prairie flora to flourish. Work continues on these plots in Manitoba, Canada.

6.5.2   Low biological potential

There are a number of ways of overcoming this problem. For vegetation the commonest approaches are to sow seeds of herbaceous and grass species, planting of trees and shrubs (cf. Chapter 5), and transplantation of turves (cf. Chapter 3). Plantrose and Plantrose (1990) have described their experiences in reintroducing wooded areas to a previously grazed island in north-west Sutherland, Isle Martin. One of the biggest problems is that of securing plants of known provenance, i.e. of local origin, and they had to make some small compromises when it came to obtaining sufficient quantities of Scots pine (*Pinus sylvestris*). Fortunately, there are increasing numbers of nurseries capable of supplying such stock. Another useful source of plant propagules is that of hay taken from species-rich meadows. In recent years this practice has become increasingly widespread and successful. Lippitt *et al.* (1994) have emphasized the importance of the care that must be taken in the various stages of collection, processing and storage of native seed. Rodwell and Patterson (1994) have given comprehensive guidance as to the establishment of new native woodlands in the United Kingdom, based on site factors and the National Vegetational Classification, including advice on the use of 'nursery crops' used to improve soil conditions specifically for the later tree and shrub dominants that it is intended to establish.

It would be hoped that once sufficient food stocks and cover had been provided then the site would be re-colonized by birds and other animals. There may be some instances where this is not possible, owing to peculiarities of geography. Diefenbach *et al.* (1993) reported on the success of a programme to reintroduce the bobcat (*Felis rufus*) to Cumberland Island, Georgia. It was established that the species had been present on the island up until the turn of the century, but had become locally extinct. A total of 32 bobcats were released on the island, captured from the nearby mainland, over a period of two years. The populations were periodically recaptured for monitoring purposes. In general, the population showed weight gain, and several litters of kittens were recorded. The annual survival rates of adults was shown to be 93 per cent, and that the population density had doubled in two years. Monitoring continues, but at this stage the reintroduction appears to be successful. The prospects for reintroduction of large predators in other systems remains a contentious one, but would appear to be feasible.

When woodlands have been established and provide some cover it becomes possible to introduce, and it is possible to bring expert systems based on decision trees to bear. Francis (1995) has described one such system, based on degree of shade (< 40 per cent >) and then abundance of existing field layer vegetation. Francis indicates that there are three types of species which may then be introduced, depending upon the outcome:

1. Woodland edge or marginal species that can be introduced as seed and are quick to mature and establish.
2. Shade-tolerant species that can be introduced as seed, but take time to establish.
3. Shade-tolerant species which are most successfully introduced as plants from which they establish predominantly from vegetative growth.

The vegetational structure then needs to be monitored to ensure that key species are reintroduced if they die out.

6.5.3 Hydrology

It may be difficult to alter the hydrology of a site without major groundworks, and the disadvantages that this brings would, in most cases, outweigh the advantages. In this case the usual approach would be to make the best of the available water table. Waterlogged areas can be drained, and dry areas at the bottom of slopes provided with a barrier to retain moisture on the side adjacent to any soak-away. More than this is difficult to achieve. At this stage species of plant should be carefully selected as to their tolerances of water-logging or droughtiness (cf. Chapter 5).

6.5.4   Substrate variability

One of the aims of agro-forestry is to eliminate the effects of variability of substrates, so it is no surprise that this is often the case on ex-farmland. Variability can be encouraged by planting in patches as outlined above, but it may also be achieved by selective fertilization, removal of substrate rich in nutrients or the importing of substrates varying in textural class and nutrient status. Jacobson *et al.* (1994) have described a project centred around the application of the concept of 'sculptured seeding'. Essentially, this aims to closely match site characteristics with plant capabilities at a level of detail of small landscape features such as knolls, dips, wet zones, dry areas and changes in soil texture. This approach will have to be taken if future restorations are to be as diverse and interesting as the originals.

6.5.5   Lack of cover

Cover is essential for most animal species. In a new woodland it is possible to install squirrel-proof nesting boxes for birds, and artificial excavations for larger animals. Importation of deadwood and rocks may also be appropriate, but in the case of the latter only when they are features of the undisturbed landscape.

6.5.6   Presence of barriers

In many cases the presence of a barrier is an ideal protection against unregulated intervention by humans, but in other cases this is not a problem. Therefore it may be necessary to remove fencing, install bridges or construct access tunnels under roadways.

6.5.7   Erodibility

This will be a problem whilst the site is not fully integrated as to its hydro-logical and ecological function, but can be addressed in the short term by use of mulches and other top dressings.

6.5.8 Lack of variability of stress and disturbance

In addition to the variability of the substrate and that imposed by climate, it may be necessary to introduce stresses and disturbances in a managed way. Mowing and grazing are two commonly applied tools in this regard, and may be used to maintain species-rich meadows. Others include the reconstitution of annual flooding events, causing physical and enrichment disturbance.

Reader and Buck (1991) described the response of the community to an experimental disturbance consisting of the creation of artificial earth mounds of three sizes designed to mimic the actions of burrowing animals on abandoned pasture. They found short-term increases in plant species richness, but this effect disappeared after two years. Unfortunately they did not extend this study by persistently changing the mound location, which may have prolonged the effect.

Smith (1995) has had considerable success in using prescribed burning techniques in restoring Oak-Hickory Woodlands and prairie grassland in the Cleveland Metropark System in Ohio. This technique suppresses undesirable invaders, and opens up the canopy to encourage an open and diverse forest architecture. Smith also indicates the central role that can be played by volunteers, in every activity from botanical survey to providing skills on the prescribed burn team. Burning techniques certainly deserve further attention and investigation, as they are low cost and do not introduce xenobiotics to the system being restored, but do require a high degree of skill and commitment from the personnel involved.

6.6 Management and after-use

At the beginning of this chapter the importance of defining the use to which the site was to be put and thus the function vegetation will have to perform was stressed. Providing that characterization of the site has been such that the appropriate treatments have been carried out to allow vegetation to function in the way desired, management is concerned with maintaining, enhancing and directing that function. Site management involves many activities in addition to management of vegetation, and like any management activity is best achieved by setting objectives, deciding strategy, implementing action plans and monitoring progress, with fine-tuning where necessary. An approach to formulating a management plan is outlined in Fig. 6.13, the types of features which may require management in Table 6.18 and the economic value of some after-uses in Fig. 6.14.

The after-uses defined will require particular vegetation types (Fig. 6.15). Table 6.19 indicates the management these vegetation types will require. However, precise management will be dictated by intensity of use, soil

RECLAMATION AIMS

A statement of the purpose of
reclamation and the broad policies
which will underlie the management of
the various land-uses and interests.

REFERENCE

A comprehensive record of the land
before and after reclamation, forming
a basis for the analysis and objectives.
External influences and constraints are
included.

ANALYSIS

An examination of the options for management of the land and the relationship
between potential land-uses. The identification of potential problems and conflicts,
and weighting of various interests. Objectives are formulated from the decisions
reached.

MANAGEMENT OBJECTIVES

Specific statement of the land-uses and interest to be achieved and promoted,
their priorities, and targets for physical/biological/financial performance.

MANAGEMENT PRESCRIPTION

An outline of the work required and the resources needed to achieve the
management objectives. A long-term outline of the programme for implementation.

IMPLEMENTATION

Details of the operations to be carried out within the review period in order to
achieve the management objectives. Statements of resource requirements related
to yearly programmes for the review period.

MONITORING AND REVIEW

An assessment and record of management achievements, with arrangements for
a periodic review of the plan and renewal or revision of the rolling implementation
programme.

Fig. 6.13 The content and operation of a management plan (source, Richards *et al.*,
1993).

Table 6.18 Features for recreational after-uses (Welsh Development Agency, 1993)

Intensive recreation		Extensive recreation	
	Required features:		**Required features:**
Sport	Flat pitches for rugby and cricket	Walking	Linear parks and routeways
	Grass and hardcourts for tennis		Extensive areas of attractive land
	Open water for sailing, canoeing		Small local areas for strolling, exercising dogs
	Swimming pools		Woodland or scrub to provide seclusion and variety
	Routes for orienteering, keep fit and cross country running	Viewpoints	High land
	Ranges for shooting and archery	Ball games	Variable topography
			Small open areas
Camping sites	Flat hard areas for caravan, car and camper standing		Landform and/or vegetation for screening
	Services, e.g. water, sewerage, electricity and site drainage		Relatively short robust ground vegetation
	Flat grass areas for pitching tents	Picnicking	Small open areas
			Landform and/or vegetation for shelter
	Tall dense vegetation to give shelter and screening	Imaginative child's play	Varied landform
	Open space for ball games and dogs		Mixture of robust vegetation types
Car-parks	Access from highway		Isolation from hazards, e.g. roads and railways
	Flat areas for vehicle parking	Organized child's play	Flat areas for swings and play equipment
	Suitable substrate		Seats for adults
	Landform and/or vegetation to screen site	Wildlife casual interest and study	Variety of habitats and associated species
	Vegetation to shade vehicles		Varied landform
Picnic sites	Flat areas for tables		Limited disturbance
	Car-parking		Pedestrian access
	Toilets	Wildlife nature conservation and scientific study	Appropriate habitats for desired species
	Tall dense vegetation for shelter and screening		Appropriate management
	Short grass to sit on		Controlled access
	Slopes to catch sun, provide shelter from wind		Minimal disturbance
	Space for ball games and dogs		

characteristics, locational factors and resources available. Other management factors which affect vegetation management are:

1. Drainage.
2. Structures.
3. Contaminated areas.
4. Areas of subsidence or settlement.
5. Litter, rubbish tipping and vandalism.
6. Public safety.
7. Public access.

What is important is that these programmes enter into the public domain as of being of 'value' and have the necessary material and economic support.

High economic value

Industry
Housing
Public buildings
and facilities
Highway improvements
Agriculture
- cultivation
- productive grazing
- marginal grazing
- allotments

Productive uses with some
economic value

Forestry
- economic
- marginal
Recreation
(intensive)
- sport
- picnic sites
- caravan and campsites
- car-parks
- waterscapes
Recreation
(extensive)
- casual/public open space
- country parks

Amenity uses with little
or no economic benefit

Wildlife
- casual/education
- nature conservation/
  education

Low economic value

Fig. 6.14 Economic value of after-uses for reclaimed land (source, Welsh Development Agency, 1993b).

6.7 Conclusions

There are now many techniques available for the establishment and mainte-nance of amenity vegetation, and our knowledge is increasing all of the time as to methods of establishing self-sustaining systems. There are still instances where failures have occurred and this is the subject of the next chapter.

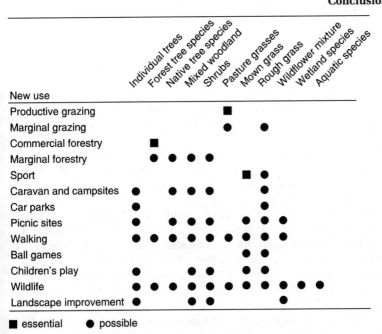

Fig. 6.15 Vegetation requirements of reclaimed land-uses (source, Welsh Development Agency, 1993b).

Table 6.19 Management requirements of vegetation on reclaimed land

Use	Summary of requirements
Wildlife	Need to ensure that the desired habitats and species develop and are sustained, e.g. by deliberate introduction of desirable species, limitation of succession to maintain seral stages such as grassland, coppicing of woodland to encourage a coppice flora.
Amenity grassland	Use may be intensive or non-intensive and management should be adapted accordingly. Considerations are: appropriate nutrient levels for balance between productivity and species diversity; timing of mowing to allow specific species to seed; appropriate use of site and control of invasive weeds; removal of thatch; control of pH; possibility of using grazing instead of mowing.
Woodland	Need for nutrients in establishment phase (not usually needed on soils) or retention of nitrogen-fixing species as a component of maturing woodland; monitoring of vigorous nurse species and selective thinning; control of fire risk (may be greater in urban situations than other sites); maintenance of fencing to exclude grazing animals; management of woodland to prevent windthrow (e.g. by coppicing).
Wetlands	Wetlands will be managed principally for wildlife and will be concerned with retaining the correct balance between open water, emergent vegetation, short turf, longer grassland and scrub. Provision and protection of nesting and roosting areas will be needed and perhaps other habitat creation/protection measures. If wetlands have a function such as the treatment of waste water then management of that function will be needed also and of the balance between maintaining treatment efficiency and wildlife value.

References

BAKER, A.J.M. and DALBY D.H. (1980). Morphological variation between some isolated populations of *Silene maritima* in the British Isles with particular reference to inland populations on metalliferous soil. *New Phytologist*, **84**, 123–138.

BAKER, A.J.M. and PROCTOR, J. (1990). The influence of cadmium, copper, lead and zinc on the distribution and evolution of metallophytes in the British Isles. *Plant Systematics Evol.*, **173**, 91–108.

BARTUSKA, A.M. and LANG, G.E. (1981). Detrital processes controlling the accumulation of forest floor litter on black locust revegetated surface mines in north central West Virginia. *Symposium on Surface Mine Hydrology, Sedimentology and Reclamation*, pp. 359–365. University of Kentucky, Lexington.

BELLAIRS, S.M. and BELL, D.T. (1993). Seed stores for restoration of species-rich shrubland vegetation following mining in Western Australia. *Restoration Ecol.*, **1**(4), 231–240.

BRADSHAW, A.D. (1993). Understanding the fundamentals of succession. In Miles, J. and Walton, D.H. (eds) *Primary succession on land*. Blackwell, Oxford.

BRADSHAW, A.D. and CHADWICK, M.J. (1980). *The restoration of land*. Blackwell, Oxford.

BRANDON, T.W. (ed.) (1989). *River engineering. Part II Structures and coastal defence works*. Number 8. IWEM, London.

BRIERLEY, J.K. (1956). Some preliminary observations on the ecology of pit heaps. *J. Ecol.*, **44**, 383–390.

CHADWICK, M.J. and HARDIMAN, K.M. (1976). Vegetating colliery spoil. Paper 27 given at the Land Reclamation Conference, Thurrock, 5–7 October 1976. Unpublished.

CHADWICK, M.J., ELIAS, C.O., LLOYD, A., MORGAN, A.L., PALMER, J.P. and WILLIAMS, P.J. (1978). *Nutrient problems in relation to vegetation establishment and maintenance on colliery spoil*. Colliery Spoil Reclamation Research Unit, University of York. Report to the Department of the Environment under contract DGR B/71.

COLBOURN, P. and STANTON, P.M. (1979). *Further work on problems in the management of soils forming on colliery spoils*. Research contract DGR 482/63 and 288-77-1 ENV UK for Department of the Environment and the European Commission. University of Newcastle upon Tyne, Newcastle upon Tyne.

COLLIS, I. and TYLDESLEY, D. (1993). *Natural assets*. Local Government Nature Conservation Initiative, Winchester.

CONNELL, H.G. and SLATYER, R.O. (1977). Mechanisms of succession in natural communities and their role in community stability and organisation. *Am. Naturalist*, **111**, 1119–1144.

COPPIN, N.J. and RICHARDS, I.G. (1990). *The use of vegetation in civil engineering*. Butterworths, London.

CROCKER, R.L. and MAJOR, J. (1955). Social development in relation to surface age at Glacier Bay, Alaska. *J. Ecol.*, **43**, 427–448.

CROXTON, W.C. (1928). Revegetation of Illinois coal stripped lands. *Ecology*, **9**, 155–175.

DAVIS, B.N.K. (1976.) Wildlife, urbanisation and industry. *Biol. Conserv.*, **10**, 249–261.

DEPARTMENT OF THE ENVIRONMENT (1994). *The reclamation and management of metalliferous mining sites*. HMSO, London.

DEPARTMENT OF THE ENVIRONMENT (in preparation). *Review of landscaping, restoration and revegation of colliery spoil heaps and lagoons*. Geology and Minerals Research Programme.

DIEFENBACH, D.R., BAKER, L.A., JAMES, W.E., WARREN, R.J and CONROY, M.J. (1993). Reintroducing bobcats to Cumberland Island, Georgia. *Restoration Ecol.*, **1**(4), 241–247.

DIXON, J.M. and HAMBLER, D.J. (1993). Wildlife and reclamation ecology: rabbit middens on seeded limestone quarry spoil. *Environ. Conserv.*, **20**(1), 65–73.

DOWN, C.G. (1973). Life form succession in plant communities on colliery waste tips. *Environ. Poll.*, **5**, 19–22.

ELIAS, C.O., MORGAN, A.L., PALMER, J.P. and CHADWICK, M.J. (1982). *The establishment, maintenance and management of vegetation on colliery spoil sites.* Department of the Environment, London.

FRANCIS, J.L. (1995). The enhancement of young plantations and new woodlands. *Land Contam. Reclam.*, **3**(2), 93–95.

FRIEDEL, H. (1938a). Die Pflanzenbesiedlung in Vorfelde des Hintereisfereners. *Zeitschrift Gletscherkunde*, **26**, 215–239.

FRIEDEL, H. (1986). Bodem und Vegetations entwicklung im Vorfelde des Rhonegletschers. *Berichte des Geobotanischen Fursch. Institutes Rübel*, Zurich 1937, pp. 65–76.

GILBERT, O.L. (1989). *The ecology of urban habitats.* Chapman and Hall, London.

GILDON, A., STANTON, P.M. and DAGLISH, P. (1982). *The use of soils in reclaiming colliery spoil.* Department of the Environment, London.

GORSIRA, B. and RISENHOOVER, K.L. (1994). An evaluation of woodland reclamation on strip-mined lands in east Texas. *Environ. Man.*, **18**(5), 787–793.

GRIME, J.P. (1979). *Plant strategies and vegetation processes.* John Wiley, Chichester.

GWENT COUNTY COUNCIL (1976). Report of the derelict land survey (draft). Unpublished.

HALL, I.G. (1957). The ecology of disused pit heaps in England. *J. Ecol.*, **45**, 699–720.

HALVORSON. J.J., SMITH, J.L. and FRANZ, E.H. (1991). Lupine influence on soil carbon, nitrogen and microbial activity in developing ecosystems at Mount St. Helens. *Oecologia*, **87**, 162–170.

HELLINGS R. (1988). *Fife derelict land review.* Nature Conservancy Council, Scottish Development Agency, Fife Regional Council.

HODGE, S.J. and HARMER, R. (1995). The creation of woodland habitats in urban and post-industrial environments. *Land Contam. Reclam.*, **3**(2), 86–88.

HOLLIDAY, R.J. and JOHNSON, M.S. (1979). The contribution of derelict mineral and industrial sites to the conservation of rare plants in the United Kingdom. *Minerals Environ.*, **1**, 1–7.

HUBY, M. (1981). *The natural colonisation of colliery spoil.* D.Phil. Thesis. University of York, York.

INGROUILLE, M.J. and SMIRNOFF, N. (1986). *Thlaspi caerulescens* J. & C. Presl (*T. alpestre* L.) in Britain. *New Phytologist*, **102**, 219–233.

IRELAND, T.T., WOLTERS, G.L. and SCHEMNITZ, S.D. (1994). Recolonization of wildlife on a coal strip-mine in northwestern New Mexico. *Southwestern Naturalist*, **39**(1), 53–57.

JACOBSON, E.T., WARK, D.B., ARNOTT, R.G., HAAS, R.J. and TOBER, D.A. (1994). Sculptured seeding: an ecological approach to revegetation. *Restoration Man. Notes*, **12**(1), 46–50.

JOHNSON, M.S. (1978). Land reclamation and botanical significance of some former mining and manufacturing sites in Britain. *Environ. Conserv.*, **5**(3), 223–228.

KELCEY, J.G. (1975). Industrial development and wildlife conservation. *Environ. Conserv.*, **2**, 99–108.

LEISMAN, G.A. (1957). A vegetation and soil chronosequence on the Mesabi iron range spoil banks, Minnesota. *Ecol. Monographs*, **27**, 221–245.

LIPPITT, L., FIDELIBUS, M.W. and BAINBRIDGE, P.A. (1994). Native seed collection, processing, and storage for revegetation projects in the Western United States. *Restoration Ecol.*, 2(2), 120–131.

MCDOUGALL, W.B. (1918). Plant succession on an artificial bare area in Illinois. *Trans. Illinois State Acad. Sci.*, 11, 129–131.

MARGULES, C.R. and USHER, M.B. (1981). Criteria used in assessing wildlife conservation potential: a review. *Biol. Conserv.*, 21, 79–109.

MARRS, R.H. and BRADSHAW, A.D. (1993) Primary succession on man-made wastes: the importance of resource acquisition. In Miles, J. and Walton, D.H. (eds) *Primary succession on land*. Blackwell, Oxford.

MARRS, R.H., ROBERTS, R.D., SKEFFINGTON, R.A. and BRADSHAW, A.D. (1983). Nitrogen and the development of ecosystems. In Lee, J.A., McNeill, S. and Rorison, I.H. (eds) *Nitrogen as an Ecological Factor*. Blackwell, Oxford.

MAUNDER, M. (1992). Plant reintroduction: an overview. *Biodiversity Conserv.*, 1, 51–61.

MOLYNEUX, J.K. (1963). Some ecological aspects of colliery waste tips around Wigan, South Lancashire. *J. Ecol.*, 51, 315–321.

MORGAN, J.P. (1994). Soil impoverishment: a little-known technique holds potential for establishing prairie. *Restoration Man. Notes*, 12(1), 55–56.

NATIONAL RIVERS AUTHORITY (1994). *The new rivers and wildlife handbook*. RSPB, Sandy.

NELSON, L.L. and ALLEN, E.B. (1993). Restoration of *Stipa pulchra* grasslands: effects of Mycorrhizae and competition from *Avena barbata*. *Restoration Ecol.*, 1(1), 40–50.

OLSEN, J.S. (1958). Rates of succession and soil changes on southern Lake Michigan sand dunes. *Botanical Gazette*, 199, 125–170.

PALMER, J.P. (1984). *An investigation of the potential for the use of legumes on colliery spoil.* PhD thesis, University of York, York.

PETIT, D. (1982). Natural vegetation of the tips of Northern France – its connection with some spoil chemical parameters. In *International Symposium on Mine Spoil Heaps*, pp. 99–118. Kommunalverband Ruhrgebiet, Essen.

PLANTROSE, B. and PLANTROSE, E. (1990). Restoring our native woodlands: a case study on the RSPB reserve of Isle Martin. *Trans. Bot. Soc. Edinburgh*, 45, 501–507.

PRACH, K. and PYSEK, P. (1994). Spontaneous establishment of woody plants in central European derelict sites and their potential for reclamation. *Restoration Ecol.*, 2(3), 190–197.

RATCLIFFE, D.A. (1974). Ecological effects of mineral exploitation in the United Kingdom and their significance to nature conservation. *Proc. R. Soc. London, A*, 339, 355–372.

READER, R.J. and BUCK, J. (1991). Community response to experimental soil disturbance in a mid-successional, abandoned pasture. *Vegetation*, 92, 151–159.

RICHARD, J.L. (1968). Les group vegetaux de la reserve d'Aletsch. *Beitraege Geobotanischen Landesaufnahme der (Schweiz)*, 51, 305.

RICHARDS, A.J. and PORTER, A.F. (1982). On the identity of a Northumberland Epipactis. *Watsonia*, 14, 121–128.

RICHARDS, A.J. and SWANN, G.A. (1976). *Epipactis leptochila* (Godfrey) Godfrey and *E. phyllanthes* G.E Sm. occurring in South Northumberland on lead and zinc soils. *Watsonia*, 11, 1–5.

RICHARDS, I.G., PALMER, J.P. and BARRATT, P.A. (1993). *The reclamation of former coal mines and steelworks*. Elsevier, Amsterdam.

RICHARDSON, J.A., SHENTON, B.K. and DICKER, R.J. (1971). Botanical studies of natural and planted vegetation on colliery spoil heaps. *Landscape Reclam.*, 1, 84–99.

RIMMER, D.L. and COLBOURN, P. (1978). *Problems in the management of soils forming on colliery spoils*. Research contract DGR 482/7 for Department of the Environment. University of Newcastle upon Tyne, Newcastle upon Tyne.

ROBERTS, R.D., MARRS, R.H., SKEFFINGTON, R.A. and BRADSHAW, A.D. (1981). Ecosystem development on naturally-colonized china clay wastes. I. Vegetation changes and overall accumulation of organic matter and nutrients. *J. Ecol.*, **69**, 153–162.

RODWELL, J. and PATTERSON, G. (1994). *Creating new native woodlands*, Forestry Bulletin 112. HMSO, London.

SCHRAMM, J.R. (1966). Plant colonisation studies of block wastes from anthracite mining in Pennsylvania. *Trans. Am. Philos. Soc.*, **56**, 1–194.

SCULLION, J. (1994). *Restoring farmland after coal: the Bryngwyn Project*. British Coal Opencast Executive, Mansfield.

SMITH, K.D. (1995). Urban forest ecosystem restoration and the role of volunteer earth restorers. *Land Contam. Reclam.*, **3**(2), 123–124.

USHER, M.B. (1986). *Wildlife conservation evaluation*. Chapman and Hall, London.

VIERECK, L.A. (1966). Plant succession and soil development on gravel outwash of the Muldrow Glacier, Alaska. *Ecol. Monographs*, **36**, 181–199.

WELSH DEVELOPMENT AGENCY (1993). *Working with nature*. WDA, Cardiff.

Further reading

FERRIS-KAAN, R. (1994). *Managing forests for biodiversity*, Forestry Commission Technical Paper 8. HMSO, London.

FERRIS-KAAN, R. (1995). *The ecology of woodland creation*. John Wiley, Chichester.

PETERKEN, G.F. (1993). *Woodland conservation and management*, 2nd Edn. Chapman and Hall, London.

Chapter 7

Why restorations fail

.7.1 Introduction

In order to determine why something has failed we have to have a concept and preferably a measure of what constitutes success. There are many cases where the aim of the restoration or reclamation programme is not achieved. This falls into two main categories:

1. Failure to implement all aspects of the restoration or reclamation plan.
2. Inadequate understanding of the ecology of the system which is the goal of the work programme.

This latter is particularly problematical as it could be the result of a simple failure of the initial monitoring programme, where an aspect of site function was overlooked, or, more fundamentally, there is inadequate understanding of the *principles of ecosystem function*. This is beyond the scope of this book, but it is interesting to note that many principles of theoretical ecology may be tested in systems subjected to the extremes found on sites in need of restoration.

With respect to the restoration of land which has been subject to the process of investigation, design, implementation, management and monitoring, the concept of success must be that the restored site and all its elements fulfil the function for which it was designed. We also have here to introduce the idea of perception of success because although a restoration may fulfil its function as far as the designer is concerned, the site may be perceived to have failed in some respect by a user or user group. The possibility that a site restoration may have been widely regarded as successful at the time of its completion and for some time afterwards and then begun to be regarded as unsuccessful has also to be considered. In this last scenario the possibilities of changes in standards, attitudes or even trends in fashion may be responsible. Perhaps here the concept of success or failure should not be used but that, as in disciplines such as building design, the idea that restoration design and function may become outdated should be entertained. If we do make this analogy then the concept that restorations can be renovated may also be suggested.

In the following sections the ways in which restorations may fail is discussed. In the discussion, attempts are made to make the distinction between the causes and symptoms of failure clear. The importance of getting the process of reclamation right and being clear about the objectives and ways of achieving those objectives is stressed.

7.2   Inadequate site surveys

The site survey or assessment made may be inadequate in a number of ways. Principally there may be a failure to take into account an overriding factor – such as the failure to take into account the presence of a contaminant of a particular type. Similarly the assessment may be too narrow as to its scope. There are many cases where only one aspect of the site has been taken into account when assessing the impact of a development resulting in degradation. A good example of this came in the 1960s and 1970s when animal populations had their lines of communication between lair and food sources severed during construction of roads, which were not replaced when the road had been built. Nowadays, tunnels and other provisions are made. Also it may be the case that only one group of animals is taken into account when assessing the site, e.g. bird life, whose needs are quite removed from terrestrial animals and plant populations.

It has already been stressed in this book that a clear idea of the function of a reclamation scheme has to be formulated before site investigation begins. The function of a scheme may be altered as a result of site investigation findings and site investigation method altered as a result of changes in objectives for the site. That such iteration occurs during the early stages of a scheme is important. However, site surveys may be inadequate and the reasons for this include:

1. Focusing on just one issue to the exclusion of others, e.g. visual impact.
2. Doing only what is necessary to meet regulatory constraints, e.g. planning permissions.
3. Concentrating on those aspects necessary to receive funding for the scheme.
4. Lack of resources.
5. Inexperienced personnel.
6. Lack of guidance on appropriate procedures for the site being investigated.

Inadequate surveys may lead to lack of identification of:

1. Features of value.

2. Opportunities for or constraints on specific remediation approaches or land-uses.
3. Management and monitoring needs.
4. User and potential user perceptions of what is needed at a site.

A framework for an integrated approach to the reclamation of a site leading to best practice is presented in Fig. 7.1. It is not the purpose of this chapter to detail all the ways in which a reclamation scheme may fail as it is obvious that a scheme may fail from the inadequate design or specification of any aspect of the scheme, whether engineering, landscape or biological. Such inadequate design may lead to inappropriate landforms, unsuitable materials for the site use, inadequate drainage, erosion and other such symptoms. The root causes of such symptoms are not adhering to frameworks such as that in Fig. 7.1 or not employing appropriate skills and experience on the project.

The biological aspects of restoration are not as easily specified as engineering aspects because we are dealing with living materials which are affected in their performance by a range of factors not capable of being predetermined (e.g. climate). For example, if concrete is made to a particular specification then, providing it is properly made, an engineer will know its precise performance criteria and tolerances. The performance of biological systems is not capable of being predicted to the same precision. There has been some standardization of biological materials, for example for agricultural and intensive amenity uses, where much is known about the performance of grass cultivars and fertilizers under controlled conditions. However, for restoration to non-intensive uses and particularly 'ecological' end-uses there has been little development of performance criteria. The remaining sections will therefore consider what goes wrong with biological systems, how this can be related to the performance of other more easily specified parts of the reclamation process and how it fits with the best practice suggestions of Fig. 7.1.

7.3   Vegetation regression

7.3.1   Causes

Apparently well-established vegetation, which initially grew vigorously and met the expectations of those involved, can progressively decline in vigour, become moribund and even die out completely. This decline is generally termed 'regression'. It can affect trees and shrubs but is most commonly seen in grassland.

The early stages of regression may appear as a yellowing of the foliage, reduced yield or growth, and stress symptoms such as early leaf fall or the onset of flowering and seed production in normally vegetative young plants. If the causes of these symptoms are uncorrected, further deterioration will occur and trees may die, or grass swards may develop bare patches. In extreme cases the vegetation may be killed completely.

Regression can occur as the result of site conditions which have deteriorated beyond the range tolerated by the vegetation, owing to lack of management or to uncontrolled chemical and physical processes in the substrate. It can also

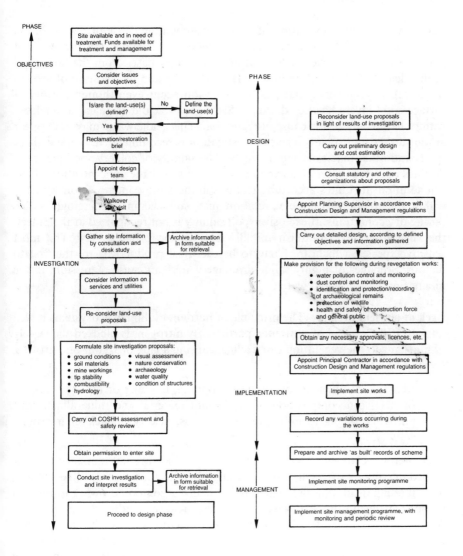

Fig. 7.1   Framework for an integrated approach to reclamation.

occur on sites where the conditions for sustained vegetation growth were not created initially and the vegetation slowly deteriorates from the outset. Excessive site use, by the public or in the form of grazing, can damage the vegetation beyond its capacity for regrowth and renewal.

7.3.2   Factors causing regression

Decline in pH   The generation of acidity is a common cause of regression on some substrates. On some colliery and metalliferous spoils the processes of pyrite oxidation and the leaching of lime and other bases from the upper soil profile, lead to the reduction of the pH and a reduction or cessation of growth (Fig. 7.2). In some cases patches of regression occur even though the pH of surface spoil layers has not declined. Seepages of acidic drainage water from within the tip may be the cause of regression. This acidic water may not appear at the surface and so sub-surface investigation is recommended. Some colliery spoils and other materials can be saline. The salinity of colliery spoil generally declines once the material is exposed to leaching, but the concentration of the soil solution fluctuates according to the soil moisture content. Spoil which is within the tolerance of the vegetation may still become damagingly saline during dry conditions. If excessive spoil salinity is not recognized at the outset, the vegetation may establish initially but decline progressively. The most sensitive species, which are likely to be the most productive, are the first to die out. Seepages of highly saline drainage water can cause regression, in the manner noted for acidic water.

Lack of nutrient cycling   The principle of nutrient cycling in ecosystems is well known. If the cycling of nutrients, particularly nitrogen, is blocked the supply of nutrients to the vegetation can decline, causing a lack of vigour and growth. Cycling can be blocked by:

1.   Removal of nutrients by grazing, cropping, leaching or fire.
2.   Ineffective decomposition of dead plant material and mineralization of nutrients, owing to excessive soil wetness, acidity or lack of microbial activity.

On many reclaimed sites, particularly those which are not managed by controlled grazing or by regular grass cutting, there is a layer of dead vegetation, and standing dead material, which is not being decomposed owing to its high carbon:nitrogen ratio, excessive soil wetness, acidity or other causes. Some of this material is ultimately blown away or otherwise lost. Such material may through its bulk prevent the establishment and growth of other plants. Material of a high carbon to nitrogen ratio will also act as a 'sink' for any incoming nitrogen (in fertilizer, via legumes or rainfall) and further restrict vegetation growth. In some sites there is no organic matter added to spoil material, which although not chemically aggressive in its own right, will lead to moribund trees.

Fig. 7.2   Vegetation regression on spoil due to pH.

Overgrazing and over-use   Grass swards, whether established for grazing or amenity uses, tolerate a degree of damage from defoliation or trampling by regrowing from the growing points of the plants. The capacity for regrowth is dependent on the species present, the supply of nutrients and other substrate conditions, and on the severity of damage sustained. Overgrazing or over-intensive use which damages the growing points of grasses and other plants, and depletes the supply of nutrients, can lead to irreversible decline in the vigour of the sward.

Decline in legumes   The value of legumes as a nitrogen source was described in Chapter 5. In agricultural and amenity grassland, if conditions for legume growth and effective nitrogen fixation are not maintained the sward as a whole will suffer from a lack of nitrogen.

Loss of soil structure   Many spoil materials are generally badly-structured, and any pore structure is usually only weakly formed. As a result, fine particles generated by weathering and those existing in the spoil are easily moved down the profile, leading to the clogging of pores. This process of consolidation is distinct from compaction due to applied loads such as tractor wheels, although the consequences are usually similar. The resulting impeded drainage and seasonal waterlogging inhibit root extension and lead to a progressively shallower root system. This in turn leads to drought stress, winter waterlogging and a decline in growth.

Deficient drainage systems   To a large extent it has been in the field of drainage technology that some of the largest advances in land restoration and reclamation practice have been made. The successful movement of water off and through substrates has been of major benefit to restoration schemes in maintaining aerobic conditions. What is clear is that in some cases drainage is left too late or is too-widely spaced, or the bore piping is too narrow.

Inappropriate materials   In some cases materials are used to cover subsoils, which are no more suitable for plant growth themselves. The failure is due to insufficient connection between the layers, low organic content, poor nutrient content, poor drainage or water-retaining characteristics, or physicochemical characteristics inappropriate to the vegetation type required as the desired end-point.

Failure of containment systems   Containment systems are vital to the sequestration of aggressive and/or toxic materials. There are numerous cases of aggressive materials corroding membranes leading to death of vegetation; a good example of this is when clay caps fail on landfill sites and leachate seeps into cover materials, giving rise to large areas of burn-off (Fig. 7.3). This occurs where the water table meets the soil surface, and is very difficult to correct.

Fig. 7.3 Leaching burn at a landfill site.

Failure of biological material   A very common cause of failure of many reclamation/restoration schemes is the failure to take into account the living nature of soil material. Too often it is treated solely as an engineering material, which may be treated in terms of its physical parameters. This approach has led to the problems associated with compaction, bulk density, waterlogging, droughtiness and the failure of plants to thrive. The introduction of more sensitive soil-handling techniques and the training of engineers as to their benefits is assisting the avoidance of such problems in the future.

Impact of off-site factors   Unfortunately there are often impacts from off-site which will compromise, and sometimes cause the failure of, restoration schemes. These can range from the arrival of nitrogen oxides in the form of atmospheric deposition, leading to eutrophication of oligotrophic systems, to plumes of acid fumes arising directly from factory chimney stacks.

Insufficient maintenance   This is probably the primary cause of the failure of restoration programmes. The maintenance may be deficient in its intensity or type. The former is due to inadequate provision of funds, the latter through inadequate (or sometimes wilful) misunderstanding of the function of the site. This may be where we reach the limits of understanding of ecosystem structure and function.

Deficiencies in legislation   There remain areas of the world where although there are no technical obstacles to the successful implementation of reclamation or restoration programmes, such programmes do not occur. This may be due to deficiencies in the legislation covering the change in land-use, and must be addressed, in the first instance, on a political level.

7.3.3   Treatment of regression

Objectives   When symptoms of regression are observed it is necessary to identify the underlying causes, and then to consider:

1. Whether the cause(s) can be rectified to enable the original vegetation objective(s) to be achieved.
2. Whether an alternative vegetation objective, which is tolerant of or avoids the cause(s), is more appropriate.

Avoidance/tolerance objectives might include acid-loving heathland vegetation in place of productive grassland dependent on lime applications, or the conversion of moribund, nutrient-poor grazing land to woodland.

Treatment of acidity   A progressive decline in pH may be halted by periodic applications of lime. In all cases with permanent vegetation, lime can only be applied to the surface and so its effect is restricted to a shallow depth, which may be inadequate to prevent regression. In arable cropping and short-term grass leys, lime may be cultivated into the substrate, and deep cultivation may be used periodically to incorporate lime before the establishment of a new crop. The cost of reseeding grasslands of low productivity or amenity value is a major obstacle to regular lime incorporation.

Nutrient cycling   The application of nitrogen to provide a steady, long-term supply maintains the productivity of a sward without stimulating unsustainable growth. The availability of nitrogen ensures that dead plant material of high carbon : nitrogen ratio can be broken down. Sewage sludge, which releases its nitrogen content over two to four years, is a valuable but inexpensive fertilizer which stimulates soil development and the development of an organic nutrient reserve. Carefully managed grazing of grassland will generally prevent the development of a 'thatch' of dead plant material, and will return nutrients and organic matter to the soil in a form which is readily recycled. Some nutrients will be lost from the sward, and so replacement is necessary.

Recultivation   Most agricultural grasslands slowly lose productivity as soil conditions deteriorate from the optimum and less productive but more persistent species become more dominant in the sward. These grasslands are renewed by incorporation of lime, fertilizers and organic matter, and reseeding. This renewal has proved successful on colliery spoil sites where regression has occurred. The technique involved:

1. Rotovation to 75 mm depth, to fragment a mat of undecomposed grass.
2. Cultivation to 300–500 mm, to disrupt the consolidated spoil profile and incorporate the organic matter from the surface.
3. Application of lime and fertilizers, according to spoil analysis.
4. Reseeding with species appropriate to the objective, e.g. productive grasses and legumes, wear-resistant sports grasses.

The renewed sward vigour which results is due to a combination of corrected pH, improved nutrient status, improved spoil structure, drainage and root penetration, and replacement of the sward with more productive species. Recultivation allows these improvements to be achieved to a much greater depth than is possible by surface applications alone. Recultivation can recreate the relatively good soil conditions which were produced initially, or allow inadequate initial ground preparation to be rectified.

Strip seeding   This technique was developed for the introduction of productive grass varieties into old pastures, without complete recultivation, but it can be used to re-establish legumes or 'wildflower' species in swards. A 'slot-seeder' cultivates narrow strips of turf and sows seed into these strips. The strip may be supplemented by a band of contact herbicide to reduce the competition for emerging seedlings. The technique will not overcome substrate conditions which are unsuitable for the introduced plants, such as acidity or phosphate deficiency, but it may prove useful once pH and nutrient status have been corrected, particularly if the site is at risk from erosion.

New site uses   Severe regression is one reason for revising the uses of the site. If severe substrate problems have developed through uncontrolled pyrite oxidation, poor landform design, severe consolidation or the failure to treat the substrate adequately at the outset, it may be better to revise the objective of the site in order to establish a more tolerant vegetation. Site problems should be treated where possible, by liming, recontouring, deep cultivation, application of sewage sludge, etc., but a less demanding vegetation type should also be considered.

External reasons for revegetation may provide the opportunity and stimulus to improve the condition of a poorly vegetated site. The current interest in woodland creation and forestry on land initially reclaimed to grazing or amenity grassland is one such reason. Flat, compacted, poorly drained sites can be shaped to provide falls for drainage and positive outlets for run-off water, and ripped to ensure deep root penetration. Soil conditions may be further improved if organic matter is added.

It is much less costly to avoid regression than to cure it. If the procedures of matching site characteristics and land-use to vegetation establishment and management described in this book are followed, regression should not occur.

7.4 Failure of aquatic systems

Aquatic ecosystems are generally easier to recreate than terrestrial ones because, providing pond or water course dimensions are suitable, water chemistry correct and there is a supply of appropriate organisms, then creation is achievable. However, such systems can fail and failure usually derives from not understanding the needs of the system to be created. This can result in, for example, eutrophication or loss of desirable species. Aquatic systems have to be managed to be successful and management may include controlling the use of the water-body or water course and its surrounding land, management of the growth of bankside vegetation (e.g. to prevent overhanging tree growth restricting light from ponds or to maintain suitable vegetation for amphibian species such as newts), desilting, removal of productive vegetation or predatory species and prevention of pollution. The same principles apply as to terrestrial vegetation in that all restoration and management activities should be geared towards fulfilling the function that the water-body or water course is intended to serve.

7.5   Objectives

The management of vegetation encompasses all the activities that are required to ensure that the vegetation continues to develop and perform the functions intended for it. All vegetation requires some degree of management, in perpetuity. Considerations of management should therefore be an integral part of the process of setting objectives and designing revegetation schemes, as these stages determine the future need for, and nature of, management. Capital inputs to revegetation works will generate minimal benefits if the vegetation declines through lack of management.

The broad aim of management efforts is usually to ensure that the original scheme objectives are fulfilled. As the site develops and is used, new land-uses and objectives may be determined. The objectives of management should therefore be reviewed at intervals, preferably by the mechanism of a management plan. The objectives should guide the day-to-day activities of management, and the interception of guidance and advice. For example, much guidance on nutrient applications is general to the production of maximum yields, whereas the manager of amenity grassland is often concerned with maintaining sward quality at minimum growth rates to minimize grass cutting, and the management of sites for floristic and wildlife interest usually requires the maintenance of low-nutrient regimes. Other objectives of management relate to the special functions that vegetation may perform on a restored site.

7.6   Management plans

A management plan can provide the site manager with the information needed to guide the consistent implementation of management policies towards the defined objectives. A plan will record the original scheme objectives for the benefit of future managers, and record any revision to the objectives determined by periodic reviews. It will also set out priorities between objectives which may at times conflict, particularly in cases of sites in multiple use. For example, in a site restored to use as a country park there might be:

1. Short-term demands to maximize the yield of grassland to generate revenue from grazing or silage.
2. Medium-term aims to develop the site's interest for visitors through species-richness and habitat diversity.
3. Long-term objectives of soil development to maximize the range of possible future land-uses.

A management plan should set out the relative priorities of these objectives, so that consistent decisions are made. Figure 7.4 gives a generalized framework for a management plan, and indicates how it may be integrated with the scheme shown in Fig. 7.1.

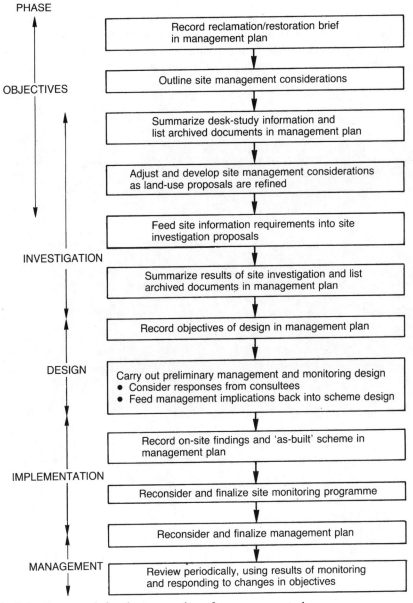

PHASE

| OBJECTIVES | Record reclamation/restoration brief in management plan |

Outline site management considerations

Summarize desk-study information and list archived documents in management plan

Adjust and develop site management considerations as land-use proposals are refined

Feed site information requirements into site investigation proposals

INVESTIGATION

Summarize results of site investigation and list archived documents in management plan

Record objectives of design in management plan

DESIGN

Carry out preliminary management and monitoring design
• Consider responses from consultees
• Feed management implications back into scheme design

Record on-site findings and 'as-built' scheme in management plan

IMPLEMENTATION

Reconsider and finalize site monitoring programme

Reconsider and finalize management plan

MANAGEMENT

Review periodically, using results of monitoring and responding to changes in objectives

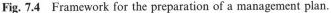

Fig. 7.4 Framework for the preparation of a management plan.

7.7   Conclusions

There remain formidable obstacles to the successful implementation of restoration and reclamation schemes, many because of the gaps in our present knowledge of ecosystem function, as opposed to failure of engineered structures, which are generally better understood. As long as we are determined to learn from our past mistakes, then future attempts at reconstruction can only be but more successful.

Chapter 8

Conclusions

8.1 Land restoration – the state of the art?

The practice and theory of land restoration and reclamation has developed significantly since the early days of 'wedding cake' reclamations of the early 1960s and the horrors of the Aberfan disaster. It has not yet, however, matured. It is still in its adolescence, where so much remains to be discovered, contemplated and finally integrated. This integration must occur on many levels. Firstly comes the integration of the professionals of different disciplines interacting with each other to produce complete (or holistic) assessments of sites, their status and potential. This includes not only scientists and engineers but planners and architects, legislators and administrators. Secondly, the process of ecosystem restoration must become rooted firmly in the public domain, so that everybody has a stake in its successful outcome, i.e. has a sense of ownership of it, such that it becomes a part of everyday life. Finally, there must come the integration of different peoples of different regions, countries and continents, with the realization that without such processes taking place, our long-term sustainability as a species may become jeopardized.

8.2 Priorities for research

A number of areas remain to be addressed on a technical level:

1. The accurate integrated measurement of ecosystem attributes, i.e. physical, biological and chemical factors, one with another and with the geology of the system.
2. Assignment of value (economic, scientific and moral) to these attributes, and the possibility of the development of absolute strictures when appropriate such as 'no species shall be made extinct' or 'no system shall be disturbed if not capable of being fully restored'.
3. The question of the importance of scale. How does the change of land-use requiring restoration in one area affect another? Is it possible to mitigate

the effects of development in one place by creating a new system in another?

4. The development of coherent and testable professional standards for both ecosystem restoration practitioners, and for ecosystem targets. Ultimately, these must include linkage between development in one region or area and compensation in another.

5. What techniques can be brought to bear on the maintenance of essential ecosystem features whilst alternative land-uses are occurring?

6. Development of comprehensive, current and interactive databases. This is where GIS has a tremendous potential for guiding the course of development and restoration, providing a means of answering many questions as to availability of genetic stocks, hydrologic function and soil resources, for example.

8.3   Priorities for legislation

This may be one area where a rearguard action fought now will reap benefit in the long term, out of proportion to the current investment. This falls largely in the development of means of enforceable and strict protection of natural resources, and the integration of legislation to include off-site considerations, i.e. setting change in use in a wider ecosystem functional context.

There also needs to be enacted legislation to bring into effect a 'ratchet of restoration'. This means that when any piece of land falls out of use then there will be a requirement that it be restored to a self-sustaining system *unless an alternative sustainable use can be immediately established*. This will prevent the inappropriate development of large areas of derelict and degraded lands. Such decisions will need to be set in the context of local, regional, national and supra-national planning. This will, of course, require a commitment from central government and the public purse, but will prove economically efficient in the long term.

8.4   Priorities for educators and activists

During the course of the preparation of this text, it has become clear that one of the principal ways forwards in securing more appropriate and effective land restoration and reclamation programmes lies in education. This may start in school, but must extend beyond that to those who believe that they have cornered the market on truth, from either side of the apparent divide. It may not always be the case that a self-sustaining use is the only permissible outcome. Perhaps the subject will have truly matured as a topic when the dedicated restorationist is able to recommend the building of a superstore on a meadow, *because it contributes towards the aim of ecosystem restoration on the larger scale*. We must listen to and learn from each other in respect to the proper and sustainable use of our land resource, and all resources that ultimately contribute to our survival and rich existence and experiences.

8.5   Prospects for restoration in the future

It is with some hope that we view the future as more and more schemes are being proposed to bolster the need to reclaim and restore in sympathy with the landscape that sustains us.

Index